CDM R
Proced

CDM Regulations Procedures Manual

Second Edition

Stuart D. Summerhayes
BSc, MSc, CEng, MICE, MaPS

Blackwell Publishing

Editorial Offices:
Osney Mead, Oxford OX2 0EL, UK
Tel: +44 (0)1865 206206
108 Cowley Road, Oxford OX4 1JF, UK
Tel: +44 (0)1865 791100
Blackwell Publishing USA, 350 Main Street, Malden, MA 02148-5018, USA
Tel: +1 781 388 8250
Iowa State Press, a Blackwell Publishing Company, 2121 State Avenue, Ames, Iowa 50014-8300, USA
Tel: +1 515 292 0140
Blackwell Munksgaard, 1 Rosenørns Allé, P.O. Box 227, DK-1502 Copenhagen V, Denmark
Tel: +45 77 33 33 33
Blackwell Publishing Asia Pty Ltd, 550 Swanston Street, Carlton South, Melbourne, Victoria 3053, Australia
Tel: +61 (0)3 9347 0300
Blackwell Verlag, Kurfürstendamm 57, 10707 Berlin, Germany
Tel: +49 (0)30 32 79 060
Blackwell Publishing, 10 rue Casimir Delavigne, 75006 Paris, France
Tel: +33 1 53 10 33 10

First edition first published 1999
Reprinted 1999, 2000, 2001
This edition published 2002

A catalogue record for this title is available from the British Library

ISBN 1-4051-0740-5

Library of Congress
Cataloging-in-Publication Data is available

Set in 11/14pt Plantin
by DP Photosetting, Aylesbury, Bucks
Printed and bound in Great Britain by
Ashford Colour Press, Gosport

For further information on
Blackwell Publishing, visit our website:
www.blackwellpublishing.com

Contents

Section 1
INTRODUCTION

Introduction

This second edition of the CDM Regulations Procedures Manual provides an opportunity to acknowledge the cultural changes that have taken place in the construction environment since the first edition was published, and also allows account to be taken of the perspective provided by the revised Approved Code of Practice and Guidance (ACoP), which came into effect on 1 February 2002. Outcomes range from, on the one hand, a general failure to improve the associated health and safety statistics, to on the other hand, the wider appreciation of health and safety management as a totally integrated sub-set of project management.

The failure of health and safety management control, as evidenced by the unacceptable level of associated statistics, indicts our processes. It is difficult to refute the criticisms, and whilst best practice health and safety management exists, it is not sufficiently widespread to suggest that the proactive and holistic approach demanded in response to the CDM Regulations is yet an integral part of the construction culture. Initiatives, however, provide optimism, and the inclusion of health and safety goals for a project, and arrangements for monitoring and review within the health and safety plan, offer opportunities that reflect industry's concerns, post-Egan, to achieve health and safety performances compatible with the ability to effectively manage the whole process.

However, these bench-marks are primarily directed at on-site management, whereas the thrust of the current ACoP is firmly directed at the positive contributions to be made by client and designer. This emphasis signifies that much still has to be done, particularly since there remain sectors of the industry and its stakeholders that still operate on a cost, not quality, driven agenda.

Significantly health and safety management has now embraced risk management concepts, with the more enlightened construction practitioners identifying the benefits of legislative response as offering far more than simply statutory compliance.

From the perspective that compliance only offers a minimum response, the best-practice operator has already identified that more stringent controls provide the more effective model. Some organisations have acknowledged that the establishment of competence represents only a starting point, and the monitoring of competence achieves a much better platform on which outcomes can be improved for all duty holders. There are many other areas where the practitioner can extend the health and safety management response for visible benefit.

Additionally, an effective CDM compliant health and safety management model provides and offers contributions towards:

- Effective project management
- Risk management
- Commercial viability
- Public relations opportunities
- Added value
- Best practice

as well as statutory compliance and the discharge of legal liabilities.

The Construction (Design and Management) Regulations 1994 are the United Kingdom's response to transpose the Council of the European Communities Directive 92/57 EEC, entitled *The Minimum Health and Safety Requirements at Temporary or Mobile Construction Sites*, into British law. It arose from member consensus that work on construction sites is not acceptably safe and that the entire responsibility and management of health and safety throughout the construction process is both fragmented and uncoordinated.

As a strategy, the CDM Regulations aim not only to reduce the unacceptable level of fatalities and major injuries associated with the construction industry but also to positively influence related aspects of occupational health and welfare.

The CDM Regulations contribute towards the cultural shift in health and safety legislation that was initiated over 20 years earlier with the launch of the Health and Safety at Work etc. Act 1974 and which signalled the move away from prescriptive legislation towards a 'deemed to satisfy' approach. The benefits of such performance-specific legislation allow health and safety issues to be managed and controlled within a risk management culture supported by a framework of quality and accountability.

The thrust of the CDM Regulations is to impose health and safety management responsibilities at all stages of a project and to extend the health and safety culture from the 'workface' of industry to the corporate management structure. The Regulations introduced an holistic approach linking all construction parties together in order to account for the health and safety management of all related issues from feasibility, through the intervening stages of design and construction inclusive of operation and maintenance up to the point of project obsolescence and the associated aspects of demolition and dismantling.

In acknowledging a changing focus over the intervening years each chapter within the manual has responded to the content of the ACoP and has sought to provide a similar degree of emphasis. This is particularly important in realising the legal status of the Code, which notes that:

> 'If you are prosecuted for breach of health and safety law, and it is proved that you did not follow the relevant provisions of the Code, you will need to show that you have complied with the law in some other way or a Court will find you at fault.'

In the discharge of construction health and safety management duties, the burden of proof rests with the duty holder. In the event of court action, the duty holder is guilty until proven otherwise.

This need not be onerous for the competent practitioner undertaking his duties in compliance with the law, nor need it create an extra burden of bureaucracy through additional paperwork. However, the scope of duties needs to be appreciated, the effective response determined, the timing of actions programmed and the trail of responsibility visible. As noted in the ACoP:

> 'CDM is intended to encourage the integration of health and safety into project management. Any paperwork should contribute to the management of health and safety.'

In summary, CDM requires:

- A realistic project programme with adequate time allowed for planning, preparation and the work itself
- Early appointment of key people
- Competent duty holders with sufficient resources to meet their legal duties
- Early identification and reduction of risks
- Provision of health and safety information from the start of the design phase, through construction and maintenance to eventual demolition, so that everyone can discharge their duties effectively
- Co-operation between duty holders

- Effort and resources proportionate to the risk and complexity of the project to be applied to managing health and safety issues

This procedures manual is intended as a contribution towards that management system control. It is not intended as an introduction to the Regulations themselves and requires some prior understanding of the interrelationship between client, designer(s), planning supervisor, principal contractor and contractor(s) as defined within the CDM Regulations. Readers requiring a more detailed introduction to the Regulations have recourse to the Approved Code of Practice, CIRIA publications, guidance notes and other texts as identified in the bibliography.

There exists no single model that represents construction industry's unilateral response to the CDM Regulations, since the non-prescriptive approach advocated allows freedom of choice. Hence, compliance is specific, but the methodology of achieving it remains judgemental. This manual therefore offers a 'control document' approach where procedures are outlined, the chronological order determined and responsibility accepted through corresponding action and dated signature, supported by proformas, standard letters, model forms and tables.

The author acknowledges that compliance with the CDM Regulations, as with all other health and safety legislation, represents a minimal response. No organisation is prohibited from establishing a more comprehensive system in pursuit of best practice and in the extension of its own proactive culture.

The success of each role, whether client, planning supervisor, designer or contractor, depends on a suitable response at the right time. Prepare adequately in advance and ensure that the role and audit trail approach is appropriate to the circumstances. Do not generate unnecessary paperwork but ensure that the health and safety management of each contract is based on good communication and appropriateness.

HOW THE MANUAL WORKS

The procedural manual is divided into sections to cover the full remit of obligations imposed on the

- Client
- Planning supervisor
- Designer(s)
- Principal contractor
- Contractor(s)

Each of these functions is qualified by a flowchart and detailed checklist. The flowchart provides a chronological route through the regulations in respect of each duty. Key to this route is the *node box* which identifies obligatory duties and procedural options together with reference to the corresponding regulation and further description, where appropriate.

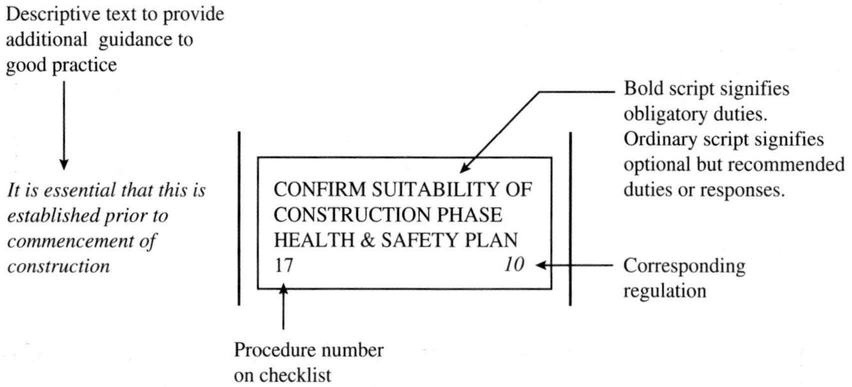

The flowchart provides an illustrated link between successive duties.

The checklist documents accountability and provides a detailed procedural step-by-step approach in fulfilment of obligations. It serves as the audit trail through the project, signifying what action needs to be taken and providing a record of what action has been taken, endorsed by the signature of the responsible person.

CONTRACT			
No.	**Reg**	**Stage**	**Procedure**
16	15(4)		• RECEIVE CONSTRUCTION PHASE HEALTH AND SAFETY PLAN

Procedure number — Corresponding regulation

Bold text script signifies obligatory duties. Ordinary script signifies optional but recommended duties or responses.

To be completed in compliance with progress

Signed and dated for accountability

	JOB REFERENCE					
Description	**Timing of action**	**Action req.**		**Initials**	**Date actioned**	**Sample document**
		Yes	**No**			
This must be forwarded by the principal contractor to the client or planning supervisor for assessment on its suitability. The planning supervisor must be available to comment if the client so wishes. Usually it is the planning supervisor who sanctions its suitability.	BEFORE COMMENCEMENT OF CONSTRUCTION					PS17

Refers to the relevant sample document to accompany action

Descriptive text to provide additional guidance to good practice

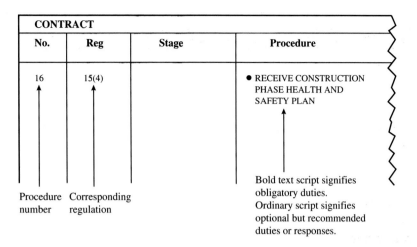

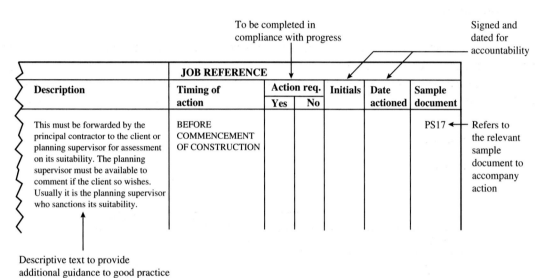

The sample documents stand as supplements consisting of a portfolio of standard letters, related questionnaires, model forms and typical proformas to complement the duties discharged by each of the functions as referenced in the associated checklist. As actioned, they provide the audit trail mechanism associated with project administration, remaining on file for the duration of the liability.

Section 2
CDM COMPLIANCE – OR NOT?

All construction projects require the effective management of health and safety. This is a prerequisite for compliance with the Health and Safety at Work etc. Act 1974 and the Management of Health and Safety at Work Regulations 1999. Other related legislation imposes more specific obligations in the discharge of duties, such as the Control of Substances Hazardous to Health Regulations 1999. Subsequently every construction activity requires appropriate risk assessments and method statements together with the provision of relevant information based on a proactive approach from all parties.

Construction projects which are subject to the CDM Regulations require an extension to the above response arising from the formalised need to establish an audit trail, evidenced by appropriate documentation and a more stringent process of accountability. However, a sense of proportion must accompany such responses since for the simplest of projects there is little more required in terms of documentation than is required in the fulfilment of other statutory duties if the project was not subject to the CDM Regulations.

Compliance with the Regulations is itself dependent on the work falling within the definition of construction and subject to the criteria in the accompanying table.

In regulation 2(1), 'construction work' means the carrying out of any building, civil engineering or engineering construction work and includes any of the following:

'(a) the construction, alteration, conversion, fitting out, commissioning, renovation, repair, upkeep, redecoration or other maintenance (including cleaning which involves the use of water or an abrasive at high pressure or the use of substances classified as corrosive or toxic for the purposes of regulation 7 of the Chemicals (Hazard Information and Packaging) Regulations 1993), decommissioning, demolition or dismantling of a structure
(b) the preparation for an intended structure, including site clearance, exploration, investigation (but not site survey) and excavation and laying or installing the foundations of the structure
(c) the assembly of prefabricated elements to form a structure or the disassembly of prefabricated elements which immediately before such disassembly formed a structure
(d) the removal of a structure or part of a structure or any product or waste resulting from demolition or dismantling of a structure or from disassembly of prefabricated elements which, immediately before each disassembly, formed a structure and
(e) the installation, commissioning, maintenance, repair or removal of mechanical, electrical, gas, compressed air, hydraulic, telecommunications, corporate or similar services which are normally fixed within or to a structure

but does not include the exploration for or extraction of mineral resources or activities preparatory thereto be carried out at a place where such exploration or extraction is carried out'.

Structure is defined as:

'(a) any building, steel or reinforced concrete structure (not being a building), railway line or siding, tramway line, dock harbour, inland navigation, tunnel, shaft, bridge, viaduct, waterworks, reservoir, pipe or pipeline (whatever, in either case, it contains or is intended to contain), cable, aqueduct, sewer, sewage works, gasholder, road, airfield, sea defence works, river works, drainage works, earthworks, lagoon, dam, wall, caisson, mast, tower, pylon, underground tank, earth retaining structure, or structure designed to preserve or alter any natural feature and any other structure similar to the foregoing or

(b) any formwork, falsework, scaffold or other structure designed or used to provide support or means of access during construction work or

(c) any fixed plant in respect of work which is installation, commissioning, de-commissioning or dismantling and where any such work involves a risk of a person falling more than 2 metres.'

Thus if work is construction by definition its status for compliance with the CDM Regulations is governed by the conditions outlined in Table 2.1.

Table 2.1 Factors affecting requirements for compliance with the CDM Regulations and/or notification to HSE.

Project particulars	Compliance with CDM	Notification to HSE
(a) Demolition/dismantling*	Yes	Only if longer than 30 days or involves more than 500 person days
(b) Project duration greater than 30 days *or* Involving more than 500 person days	Yes	Yes
(c) More than 4 people on site at any one time including supervision	Yes	Yes, provided that (b) applies
(d) Fixed term maintenance ** contracts where at any one time no more than 4 people are employed including supervision	No	No

* Definition or dismantling is defined as 'The deliberate pulling down, destruction or taking apart of a structure, or a substantial part of a structure. It includes dismantling for re-erection or re-use. Demolition does not include operations such as making openings for doors, windows or services or removing non-structural elements such as cladding, roof tiles or scaffolding.' Where the rest of the work would not be CDM-related then only the demolition/dismantling would need to comply.

** Maintenance contracts as defined within CDM do not include day to day maintenance work.

Note:

- All projects, irrespective of size or complexity, need an effective approach to the management of risk. This is achieved by appropriate risk assessments and method statements. Such requirements are necessary regardless of CDM project status or otherwise.

Section 3
THE CLIENT

The client

The client fulfils the key function within the construction environment, since not only does he or she enter into contract with the main players but also exercises financial control over the project in terms of funding and payment.

The client's role is therefore instrumental in appointing competent parties who are adequately resourced. These twin issues of competence and resourcing are fundamental to the discharge of the client's duties and are directed at health and safety management. Technical and financial implications are outside the scope of the regulations and although these issues are judged elsewhere in project management their impact should not be ignored in respect of health and safety since a technically incompetent or financially unsound organisation is unlikely to adequately resource health and safety.

In order to address the issues for which clients are legally responsible the ACoP advises that the client's arrangements should ensure:

- Allowance of adequate time for design, planning, preparation and construction work
- All parties are competent and adequately resourced in health and safety matters
- Implications for overlaps involving public safety or existing occupancy are properly addressed
- Legal responsibilities and inter-relationships are clear
- Hazard identification and control measures are appropriate in accordance with Regulation 13 duties of the designer(s) in respect of the CDM Regulations and Regulation 3 duties of contractors in respect of the Management Regulations and related legislation
- Systematic and routine monitoring and review is undertaken to ensure that it is undertaken safely and without risk to health
- Revisions to designs, programmes of work or method statements are managed safely and without risk to health

Clients must adopt a proactive and reasonable approach to the provision of information. The discovery of hazards arising from the historical legacy of the site must not be left to the contractor to discover.

Clients should also realise that their obligation to ensure all information 'about the state or condition of any premises' is available (regulation 11(1)) obviates against the use of bland statements such as...

> 'the possibility exists that there may be asbestos...'
> 'the possibility exists that the land might be contaminated...'
> 'there is the possibility that there might be underground services...'

The client is expected to make efforts to provide that information based on a reasonably proactive approach. Vital up-front information must be provided by the client based on reasonable enquiries and proactive investigations, so that tendering contractors can provide suitable and sufficient resources for the work to be undertaken safely, based on information provided.

Regulation 2(1) defines the client as:

> 'any person for whom a project is carried out, whether it is carried out by another person or is carried out in-house.'

The identity of the client, as with other duty holders needs to be established as early as possible. In complex situations this can be difficult and the ACoP offers the following guidance in that the factors to be accounted for include:

- who heads the procurement chain
- who arranges for design work
- who engages contractors

Such duties can easily fall as obligations to be undertaken by other duty holders, e.g.:

- a contractor entering into a design contract
- a project manager appointing contractors

and reflect the procurement strategy of the project.

The client may appoint an agent to act on his or her behalf subject to the terms of regulation 4(2) which stipulates that:

> 'no client shall appoint any person as his agent under paragraph 1 unless the client is reasonably satisfied that the person he intends to appoint as his agent has the competence to perform the duties imposed on a client by these Regulations.'

The term agent is further defined as:

> 'any person who acts as agent for a client in connection with the carrying on by the person of a trade, business or other undertaking (whether for profit or not).'

The appointment of a competent agent undertaking the client's duties requires a written declaration to be sent to the Health and Safety Executive.

The service agreement between the client and the planning supervisor dictates the level of service provided by the planning supervisor (see Section 4, 'The Planning Supervisor'). However whilst the planning supervisor must be, 'in a position to give adequate advice', statutory duties cannot be transferred and the client needs to be aware that he/she remains responsible for:

- The provision of relevant information (regulation 11(1))
- The establishment of competence (regulation (8))
- The establishment of the adequacy of resources (regulation (9))
- The start of the construction phase (regulation (10))

A major control mechanism exercised by the client remains the sanctioning of the suitability of the construction phase health and safety plan. This represents a critical control as acknowledged by the law in that non-compliance with regulation 10 (*start of the construction phase*) represents one of only two breaches of duty, which confer a right of action in any civil proceedings. The other being with the principal contractor's duty for the security of the site (regulation 16(1)(c)).

Ultimately, a health and safety file will be provided by the planning supervisor to facilitate the future health and safety management aspects of the completed project.

The client must therefore give consideration to the health and safety file in terms of:

- Format [contents (see sample document PS 21)]

- Form, e.g. lever arch, CD ROM, microfiche, electronic
- Retrieval systems (access, indexing, abstracts, possession)
- Currency (health and safety file must be updated as modifications and alterations occur to the asset)

The client must also consider the legal documentation to accompany the partial handover of any site. Incoming tenants require relevant abstracts of the health and safety file and relevant information must accompany such contractual arrangements. There is no difficulty in handing over a partially completed health and safety file with remaining information to be provided accordingly. As such this requires discussion between the client and his/her planning supervisor, since the latter is charged with handing over the health and safety file to the client under his remit in compliance with regulation 14.

Domestic clients are exempt from compliance with the CDM Regulations.

> 'Furthermore a domestic client "means a client for whom a project is carried out not being a project carried out in connection with the carrying on by the client of a trade, business or other undertaking (whether for profit or not)".'

Domestic clients involved in the building of an extension for commercial purposes (i.e. an office, etc.) to their homes would lose their domestic client status.

Additional guidance provided in HSG 224 clarifies the situation where clients engage parties based outside Great Britain:

> 'People working abroad have no duties under CDM, but the law still applies to clients in Britain who choose to use designers or others working abroad. These clients must make sure that those they engage are competent; that health and safety issues are properly considered; and that the normal CDM information is provided. This can be covered in contracts.'

This has often proved a difficult area since designers based outside Great Britain are usually employed because of particular skills or from past associations. Subsequently, their involvement is critical to the success of the project and such criticality can dictate difficulty in ensuring their compliance with the CDM Regulations. The law however makes no distinction and clients should note that the entire construction process from the conceptual stage through to obsolescence and demolition/dismantling is subject to the regulations. There is a co-ordinating role for both the lead designer and planning supervisor in this respect.

3.1
CLIENT FLOWCHART

The following procedures are required for all projects having CDM status.

Feasibility/design	Design	Pre-construction	Construction	Post-construction
APPOINT AGENT • ESTABLISH COMPETENCE 1 *Optional*	A client may appoint an agent provided that agent is competent to perform the duties of the client			
APPOINT PLANNING SUPERVISOR **• ESTABLISH COMPETENCE** **• ENSURE ALLOCATION OF ADEQUATE RESOURCES** 2 *6(1)(a)*	Competence and adequate resourcing are essential prerequisites for successful health and safety managment			
CONFIRM EXTENT OF PLANNING SUPERVISOR SERVICE 3	This is critical so that both client and planning supervisor are aware of the extent of service to be provided			
RECEIVE COPY OF INITIAL HSE NOTIFICATION 4	The initial notification to the Health and Safety Executive should be copied by the planning supervisor to the client			
PROVIDE RELEVANT INFORMATION ON STATE AND/OR CONDITION OF PREMISES 5 *11(1) & 11(2)*		The client must be proactive in providing such information, based on reasonable enquiries having been made		
APPOINT DESIGNER(S) **• ESTABLISH COMPETENCE** **• ENSURE ALLOCATION OF ADEQUATE RESOURCES** 6 *8(2) & 9(2)*	The competence and allocation of adequate resourcing can be undertaken by the planning supervisor if the client wishes			

Feasibility/design	**Design**	**Pre-construction**	**Construction**	**Post-construction**
		APPOINT PRINCIPAL CONTRACTOR • **ESTABLISH COMPETENCE** • **ENSURE ALLOCATION OF ADEQUATE RESOURCES** 7 *8(3) & 9(3)*	Whilst the competence and allocation of adequate resourcing must be established by the client, the planning supervisor shall be in a position to give adequate advice to facilitate this process.	
		RECEIVE ADDITIONAL HSE NOTIFICATION (F10) 8	This should be provided by the planning supervisor	
	This is the critical control check and irrespective of the planning supervisor facilitating this process it remains a client statutory duty to ensure suitability. **THIS MUST BE DONE BEFORE START OF CONSTRUCTION**	**ENSURE SUITABILITY OF CONSTRUCTION PHASE HEALTH AND SAFETY PLAN** 9 *10*		
			Delivered by the planning supervisor	RECEIVE HEALTH AND SAFETY FILE AND CONFIRM RECEIPT 10
			Availability must be maintained for any person requiring such information The client needs to retain possession of the health and safety file Future modifications, etc. require updates to the health and safety file	**MAINTAIN AVAILABILITY AND CURRENCY OF HEALTH AND SAFETY FILE** 11 *12(1)*
			Change of ownership requires transfer of the health and safety file	**TRANSFER HEALTH AND SAFETY FILE WITH OWNERSHIP** 12 *12(2)*

3.2
CLIENT CHECKLIST

The following procedures are requirements for all projects having CDM status.

CONTRACT					**JOB REFERENCE**					
No.	**Reg**	**Stage**	**Procedure**	**Description**	**Timing of action**	**Action Req**		**Initials**	**Date actioned**	**Sample document**
						Yes	**No**			
1	*4(1)* *4(2)*	Feasibility/ design	APPOINT AGENT (IF REQUIRED) • ESTABLISH COMPETENCE	The client must be satisfied that the person appointed is competent	On commencement of project					CLIENT 1
2	*6(1)(a)*	Feasibility/ design	**APPOINT PLANNING SUPERVISOR**	The client must appoint a planning supervisor for every CDM-related project, subject to competence and resourcing being established	Immediately a project has been identified as CDM-related					
	8(1)		• **ESTABLISH COMPETENCE**	The client must be satisfied that the person/organisation appointed has the necessary competence to perform the function of planning supervisor	Prior to appointment					CLIENT 2
	9(1)		• **ENSURE ALLOCATION OF RESOURCES**	Similarly, there is the need to ensure that adequate resources are allocated to perform the function of planning supervisor	Prior to appointment					CLIENT 2
3		Feasibility/ design	CONFIRM EXTENT OF PLANNING SUPERVISOR SERVICE TO BE PROVIDED	The scope of the duties needs to be determined since on occasions the client may wish for a more comprehensive approach to be adopted	Prior to the appointment of the planning supervisor					CLIENT 3
4		Design	RECEIVE COPY OF INITIAL HSE NOTIFICATION (F10)	This is the first stage notification to be sent to the Health and Safety Executive giving outline information on the project. Whilst the planning supervisor only has to ensure this notification is sent to the HSE, it is usually he/she who does it	As soon as possible after the planning supervisor has been appointed					PS1
5	*11(1)* *11(2)*	Design	**PROVIDE RELEVANT INFORMATION ON STATE AND/OR CONDITION OF PREMISES**	The client must be proactive in making reasonable enquiries to provide information for the planning supervisor in respect of the premises or project generally, e.g. Asbestos Contamination Existing drawings Instability Underground services It is not sufficient to draw attention to the possibility that asbestos or contamination exists – *find out?*	As soon as possible but before the pre-construction phase health and safety plan is drafted					

CONTRACT					JOB REFERENCE					
No.	**Reg**	**Stage**	**Procedure**	**Description**	**Timing of action**	**Action Req**		**Initials**	**Date actioned**	**Sample document**
						Yes	**No**			
5 *Contd.*				*(Such information may also be relevant to the design team)* **THE PROVISION OF RELEVANT INFORMATION IS BASED ON REASONABLE ENQUIRIES HAVING BEEN MADE**						
6		Design	APPOINT DESIGNER(S)							
	8(2)		• **ESTABLISH COMPETENCE**[1]	Competence must be established. This can be done via 'Approved List' procedures or in some cases in advance of the individual design contract	Prior to appointment					PS2
	9(2)		• **ENSURE ALLOCATION OF ADEQUATE RESOURCES**[2]	Similarly, there is the need to ensure that adequate resources are allocated to perform the design function *The planning supervisor shall be in a position to give adequate advice on both competence and resource establishment and can therefore facilitate both processes.* FOR IN-HOUSE APPOINTMENTS, ALTHOUGH AN APPROVED LIST WOULD NOT BE DRAWN UP THE CLIENT STILL NEEDS TO ENSURE COMPETENCE AND RESOURCE LEVELS ARE ADEQUATE						PS2
7	*6(1)(b)*	Post-tender pre-construction	**APPOINT PRINCIPAL CONTRACTOR**	The client must appoint a principal contractor for every CDM-related project to manage the construction phase of the project, subject to competence and resourcing being established	Simultaneously with the letting of the contract					
	8(3)		• **ESTABLISH COMPETENCE**[1]	The client must be satisfied that the organisation appointed has the competence to perform the function of principal contractor. This is often established via pre-qualification procedures, e.g. Approved List	Prior to invitation to tender This could be undertaken by the planning supervisor if so required					PS7, PS9

[1] Competence is organisation-orientated and backward-looking.
[2] Resource adequacy is project-specific and forward-looking.

CONTRACT					JOB REFERENCE					
No.	**Reg**	**Stage**	**Procedure**	**Description**	**Timing of action**	**Action Req**		**Initials**	**Date actioned**	**Sample document**
						Yes	**No**			
7 *Contd.*	*9(3)*		• **ENSURE ALLOCATION OF ADEQUATE RESOURCES**[2]	Similarly, the allocation of adequate resources must be established. This can be undertaken by the planning supervisor via an interview session or other means	Prior to letting the contract					PS10, PS11, PS12, PS13
8		Pre-construction	RECEIVE ADDITIONAL HSE NOTIFICATION (F10)	This is the final stage notification sent to the HSE by the planning supervisor giving further detailed information about the project. Whilst the planning supervisor only has to ensure notification is sent to the HSE, the planning supervisor usually undertakes this duty	Immediately after the appointment of the principal contractor and **BEFORE COMMENCEMENT OF THE CONTRACT**					PS15
9	*10*	Pre-construction	**ENSURE SUITABILITY OF CONSTRUCTION PHASE HEALTH AND SAFETY PLAN**	The client must be satisfied that the principal contractor has planned the contract satisfactorily in terms of health and safety management. Thus the construction phase health and safety plan must be deemed suitable **BEFORE COMMENCEMENT OF CONSTRUCTION**. This process can be facilitated by the planning supervisor who 'shall be in a position to give adequate advice to the client'.	PRIOR TO COMMENCEMENT OF CONSTRUCTION **FAILURE TO ENSURE COMPLIANCE WITH REG 10 CONFERS AN IMMEDIATE CIVIL RIGHT OF ACTION**					PS18
10	*14(f)*	Project completion	HEALTH AND SAFETY FILE	This is the health and safety manual for the project and is handed over to the client by the planning supervisor. It remains the property of the client thereafter	On completion of the project (PREFERABLY AT PROJECT HANDOVER)					PS20
			CONFIRM RECEIPT OF HEALTH AND SAFETY FILE	Confirmation of receipt and acceptance of the health and safety file should be relayed to the planning supervisor. This effectively completes the planning supervisor's involvement in the project	Once the client is satisfied with the contents of the health and safety file					PS20
11	*12(1)*	Post-construction operation and maintenance	**MAINTAIN AVAILABILITY AND CURRENCY OF HEALTH AND SAFETY FILE**	This is an ongoing client's duty and the information within the health and safety file must be kept up to date and available for inspection by any person who may need that information	Throughout life of project					
12	*12(2)*	Change of ownership	**TRANSFER HEALTH AND SAFETY FILE WITH OWNERSHIP**	The health and safety file must accompany change of ownership	Simultaneously with the conveyancing process					

[2] Resource adequacy is project-specific and forward-looking.

Section 4
THE PLANNING SUPERVISOR

The planning supervisor

The planning supervisor is in effect *'a creature of the regulations'*, since this role is created solely by the regulations.

The function of the planning supervisor is to facilitate and coordinate the health and safety management process with particular emphasis on the design and planning phases. It does not translate into that of being the:

(1) TECHNICAL AUDITOR, or the
(2) SAFETY ADVISOR

Contractually the planning supervisor has limited powers, deriving all his facilitation from protocols that are established between the various parties.

The regulations require that the planning supervisor shall be in a position to give adequate advice to both client and contractor in the establishment of competence and the adequacy of resourcing, and for the confirmation of the suitability of the construction phase health and safety plan if so required. This availability in terms of the designer competence infers that the planning supervisor is the first appointment prior to that of the architect or consulting engineer. This is in contrast to the historical links that exist between designer and client, and subsequently the planning supervisor is often appointed at a later stage to the detriment of the process.

The integration of the planning supervisor as a team player is often undermined by his/her belated appointment, which removes the opportunity of proactive involvement and therefore undermines much added value that can emanate from the effective execution of the role.

It should be noted that there is a wide range between the planning supervisor's discharge of duties in the minimal sense and the full service that could be provided dependent upon the client's requirements.

Much of the planning supervisor's duties rests on *ensuring* that:

- Notification has been given to the HSE
- Designers are fulfilling their design risk assessment strategy (regulation 13(2)) duties
- A health and safety file is prepared
- The health and safety file is delivered to the client

There is also the need for the planning supervisor to be in a position to give adequate advice to:

- Both client and contractor in respect of the establishment of competence and adequacy of resourcing in the appointment of the designer
- The client in the establishment of competence and adequacy of resourcing in the appointment of the contractor
- The client in facilitating the sanctioning of the construction phase health and safety plan

These are important issues, which if not fully addressed could compromise the legal position of both client and planning supervisor through failure to discharge all relevant issues.

The service agreement between client and planning supervisor should be agreed at the earliest opportunity and guidance is provided through the standard Forms of Appointment provided by the Association of Planning Supervisors, Institution of Civil Engineers, Royal Institute of British Architects and Royal Institute of Chartered Surveyors.

The planning supervisor should remember that in being positioned 'to give adequate advice', he/she is merely facilitating. The statutory duty remains with the other parties.

Regulation 14(b) requires the planning supervisor to ensure co-operation between designers, in order to enable them to comply with their health and safety management duties. This requirement extends to the sub-design interface, which can present a real challenge since much sub-design is undertaken in the latter stages of a construction programme. This is often a frenetic phase of construction, with productivity and handover targets being driven through a diminishing window of opportunity. Nevertheless duties remain to be discharged.

A recurring control failure surrounds the handover stage of the health and safety file. Regulation 14(f) requires that the health and safety file is delivered to the client 'on the completion of construction work on each structure'. The provision of relevant information is often delayed and subsequently the handover becomes belated. Partial handover of relevant information remains a possibility particularly with an incoming tenant/occupier as part of sectional or substantial completion, but the full package represents the objective for compliance. The protocols for achieving this receipt of information should be formulated before the construction phase starts and the communication routes should be accessible throughout so that requisite information is provided at the appropriate time.

The planning supervisor is responsible for the content but not the detail of the information within this document and should be alert to the receipt of relevant information compatible with the construction and design programmes.

The planning supervisor should note that whilst the role ensures a health and safety file is prepared and handed over, it also requires that the planning supervisor shall 'review, amend or add to the health and safety file ... as necessary' to ensure that it contains the relevant information. Thus the planning supervisor remains responsible for the compatibility of its contents.

The role of planning supervisor can be successfully undertaken by existing construction professionals and can be fulfilled by architects, consulting engineers, project managers, civil engineers, quantity surveyors, building services engineers and health and safety professionals. The author's experience suggests that the competent and adequately resourced planning supervisor requires at least an appreciation of the procurement process, design methodology, construction methodology and health and safety legislation, together with the necessary interpersonal skills to communicate with all parties, regardless of his or her construction pedigree.

4.1
PLANNING SUPERVISOR FLOWCHART

Feasibility/design	Design	Pre-tender	Construction
ESTABLISH EXTENT OF SERVICE TO BE PROVIDED 1	This is critical to achieve client satisfaction and to establish a benchmark for comparing services prior to appointment under fee competition		
ENSURE INITIAL NOTIFICATION TO HSE (F10). COPY TO CLIENT 2 *7(1)*	The planning supervisor has only to **ensure** this is done. It is, however, usually undertaken by the planning supervisor		
CLIENT TO FORWARD RELEVANT INFORMATION ON STATE AND/OR CONDITION OF PREMISES 3 *11(1)*	This is required as part of the design coordination and will feed into the design process and the pre-tender health and safety plan and possibly the health and safety file		
ESTABLISH DESIGNER(S) • COMPETENCE • ALLOCATION OF ADEQUATE RESOURCES 4 *8(2) & 9(2)*	**The planning supervisor shall be in a position to give adequate advice to both the client and contractor on both competence and resourcing**		
CONFIRM DESIGNER(S) • COMPETENCE • ALLOCATION OF ADEQUATE RESOURCING TO CLIENT 5 *8(2) & 9(2)*	**ENSURE DESIGN CONSIDERS:** **• AVOIDANCE** **• PREVENTION** **• PROTECTION** 6 *14(a)(i)*	This is based on the hierarchy of design response, namely: • ELIMINATION • REDUCTION • TRANSFER	
	ENSURE DESIGN HAS INCLUDED ADEQUATE INFORMATION ABOUT: **• PROJECT** **• STRUCTURE** **• MATERIALS** 7 *14(a)(ii)*	Reasonable enquiries must be made by client, planning supervisor and designers to ensure relevant information is provided for the purpose of health and safety management. This may well have implications for the designer in respect of the pre-tender health and safety plan and/or the health and safety file	
	ENSURE COOPERATION BETWEEN DESIGNER(S) 8 *14(b)*	This requires proactivity by the planning supervisor. Sub-design elements within projects are also subject to a similar process of health and safety cooperation	
	RECEIVE RISK ASSESSMENT[1] PROFORMAS FROM DESIGNER(S) 9	Useful facility for the communication of residual or transferable hazards. Some of this information will eventually find its way into the pre-tender health and safety plan and/or health and safety file	

[1] Designer risk assessment proformas are a primary means of ensuring compliance with Regulation 13(2). They also provide a communication link for notification of residual risks.

Feasibility/design	Design	Pre-tender	Construction
	ENSURE PREPARATION OF PRE-TENDER HEALTH AND SAFETY PLAN 10 *15(1)*	Although the planning supervisor has only to *ensure* this is done the planning supervisor usually undertakes its preparation. This document is sent out to tendering contractors	
	ESTABLISH CONTRACTOR(S) COMPETENCE 11 *8(3)*	If required by client.	
	CONFIRM TO CLIENT CONTRACTOR COMPETENCE 12	INTERVIEW SESSION TO UNDERTAKE ALLOCATION OF ADEQUATE RESOURCES 13 *9(3)*	The allocation of adequate resourcing can be undertaken by methods other than interview, e.g. • Schedule of rates • Resource breakdown • Bill of quantity item, etc. *INTERVIEW SESSION RECOMMENDED*
		CONFIRM CONTRACTOR'S ALLOCATION OF ADEQUATE RESOURCES 14	
		ENSURE ADDITIONAL NOTIFICATION FORWARDED TO HSE (F10). THIS REQUIRES • PRINCIPAL CONTRACTOR'S SIGNATURE • COPY TO CLIENT 15 *7(4)*	Planning supervisor has only to *ensure* that this is done, but usually undertakes the notification. Important to obtain principal contractor's signature since this is the only document where the principal contractor signs to that effect
		RECEIVE CONSTRUCTION PHASE HEALTH AND SAFETY PLAN 16 *15(4)*	**The planning supervisor shall be in a position to give adequate advice on the suitability of this document if required by the client.**
	It is essential that this is established prior to commencement of construction	CONFIRM SUITABILITY OF CONSTRUCTION PHASE HEALTH AND SAFETY PLAN TO CLIENT 17 *10*	**The construction phase health and safety plan must be sanctioned by the client prior to commencement of construction.**
			ENSURE HEALTH AND[2] SAFETY MANAGEMENT IS ADDRESSED AT PROGRESS MEETINGS 18

[2] An integrated agenda item for health and safety management should be addressed at each progress meeting (refer to sample document PS19).

Feasibility/design	Design	Pre-tender	Construction
			RECEIVE INFORMATION FOR HEALTH AND SAFETY FILE FROM • CLIENT • DESIGNER(S) • PRINCIPAL CONTRACTOR **PREPARE AND COLLATE HEALTH AND SAFETY FILE** 19 *14(d)*
		This effectively determines the fulfilment of planning supervisor duties once its contents have been accepted	**DELIVER HEALTH AND SAFETY FILE TO CLIENT** 20 *14(f)*
			CONFIRMATION FROM CLIENT OF RECEIPT AND ACCEPTANCE OF HEALTH AND SAFETY FILE 21

4.2
PLANNING SUPERVISOR CHECKLIST

CONTRACT					**JOB REFERENCE**					
No.	**Reg**	**Stage**	**Procedure**	**Description**	**Timing of action**	**Action Req**		**Initials**	**Date actioned**	**Sample document**
						Yes	**No**			
1		Feasibility/design	ESTABLISH EXTENT OF SERVICE TO BE PROVIDED	This represents the agreed service provided for the client by the planning supervisor	Before appointment					CLIENT 3
2	*7(1)*	Design	**ENSURE INITIAL F10 NOTIFICATION TO THE HEALTH AND SAFETY EXECUTIVE**	This must be sent to the Health and Safety Executive giving outline information about the project	Immediately after appointment					PS1
			• COPY TO BE FORWARDED TO CLIENT	For client's record purposes						
3	*11(1)*	Design	RECEIVE RELEVANT INFORMATION ON STATE AND/OR CONDITION OF PREMISES FROM CLIENT	The client must forward relevant information corresponding to the premises or project to the planning supervisor. This will help to highlight the related health and safety management issues, e.g. • Asbestos • Contamination • Existing drawings • Instability *(SUCH INFORMATION MAY ALSO BE RELEVANT TO THE DESIGN TEAM)*	As soon as possible after appointment but before the pre-tender health and safety plan is drafted					
4	*8(2)*	Design	ESTABLISH DESIGNER'S • COMPETENCE[1]	**The planning supervisor shall be in a position to give advice on the establishment of the designer's competence, if required by the client or contractor.**	Prior to appointment of designer(s)					PS2
	9(2)		• ALLOCATION OF ADEQUATE RESOURCES[2]	**Similarly, the planning supervisor shall be in a position to give advice on the designer's allocation of adequate resources for health and safety management if required by the client or contractor.**	Prior to appointment of designer(s)					PS2
5			CONFIRM DESIGNER'S COMPETENCE AND ALLOCATION OF ADEQUATE RESOURCES	To be forwarded to the client after establishment	Prior to appointment of designer(s)					PS4

[1] Competence is organisation-orientated and backward-looking.
[2] Resource adequacy is project-specific and forward-looking.

CONTRACT					JOB REFERENCE					
No.	**Reg**	**Stage**	**Procedure**	**Description**	**Timing of action**	**Action Req**		**Initials**	**Date actioned**	**Sample document**
						Yes	**No**			
6	*14(a)(i)*	Design	**ENSURE DESIGN CONSIDERATIONS HAVE GIVEN ADEQUATE REGARD TO:** • **AVOIDANCE OF FORESEEABLE RISKS** • **COMBATING RISKS AT SOURCE** • **PRIORITISATION OF MEASURES TO PROTECT ALL PERSONS AT WORK**	The planning supervisor must ensure that a risk assessment strategy is being followed by the design team and that a design contribution is being made to health and safety management based on the hierarchical response of: • Elimination • Reduction • Transfer	Throughout design					
7	*14(a)(ii)*	Design	**ENSURE DESIGN HAS INCLUDED ADEQUATE INFORMATION ABOUT** • **PROJECT** • **STRUCTURE** • **MATERIALS**	Alternatively, notification of residual hazards can be communicated separately for inclusion in the pre-tender health and safety plan and/or health and safety file	Throughout the design process, but such information is essential before the pre-tender phase health and safety plan is completed					
8	*14(b)*	Design	**ENSURE COOPERATION BETWEEN DESIGNERS SO THAT ADEQUATE REGARD IS GIVEN TO ASPECTS OF HEALTH AND SAFETY MANAGEMENT**[3]	The coordination of health and safety issues relevant to design must be undertaken by the planning supervisor, who must ensure that information flows between the designers and that relevant health and safety issues are fully considered by everyone involved in the design process	Throughout the design process, inclusive of sub-design, via suitable protocols that the planning supervisor has established, e.g. design/planning supervision, coordination meetings, etc.					
9		Design	RECEIVE RISK ASSESSMENT PROFORMAS FROM DESIGNER(S)	This provides documentary evidence that a risk assessment strategy has been initiated, and provides information on the RESIDUAL HAZARDS that have to be transferred into the pre-tender health and safety plan and possibly the health and safety file	Prior to drafting of the pre-tender phase health and safety plan					PS5
10	*15(1)*	Pre-tender	**ENSURE PREPARATION OF PRE-TENDER HEALTH AND SAFETY PLAN**	This is a key document and notifies the tendering contractors of the significant health and safety issues specific to the individual project. The planning supervisor usually prepares it but legally has only to *ensure* its preparation	To accompany the tender documentation forwarded to tendering contractors					PS6 (sample)

[3] For design and build procurement strategies, the designer/planning supervisor coordination process continues further into the construction phase.

CONTRACT					**JOB REFERENCE**					
No.	**Reg**	**Stage**	**Procedure**	**Description**	**Timing of action**	**Action Req**		**Initials**	**Date actioned**	**Sample document**
						Yes	**No**			
11	*8(3)*	Pre-tender	ESTABLISH CONTRACTOR(S) COMPETENCE	**The planning supervisor shall be in position to give adequate advice on the contractor's competence if required by the client.** This is often addressed during pre-qualification procedures for acceptance onto an APPROVED LIST	Prior to invite to tender					PS7, PS8
12			CONFIRM ESTABLISHMENT OF CONTRACTOR COMPETENCE	Confirmation is to be forwarded to the client once competence has been established	Prior to invite to tender					PS9
13		Post-tender pre-contract	ENSURE ALLOCATION OF ADEQUATE RESOURCES	**The planning supervisor shall be in a position to give adequate advice on the adequacy of resourcing if requested by the client** It is recommended that this is established via an interview. Minutes of such an interview must be forwarded to the favoured contractor for perusal, amendment and signature and returned to the planning supervisor. Other methods may also be adopted.	After tender scrutinisation but BEFORE THE CONTRACT IS LET					PS10 (sample), PS11 (sample), PS12
14			CONFIRM CONTRACTOR'S ALLOCATION OF ADEQUATE RESOURCES	Confirmation is to be forwarded to the client after receipt of signed minutes from favoured contractor	Prior to letting the contract					PS13
15	*7(4)*	Post-tender pre-construction	**ENSURE ADDITIONAL F10 NOTIFICATION IS FORWARDED TO THE HSE**	This must be forwarded to the HSE after further information has been added to the original form and **requires the signature of the principal contractor**	Immediately the contract is let and **BEFORE COMMENCEMENT OF CONSTRUCTION WORK.** Best done in the pre-letting meeting					PS14, PS15
			• COPY TO PRINCIPAL CONTRACTOR	**For the principal contractor to display on site in a prominent position**	Simultaneously with above					PS16
			• COPY TO CLIENT	For client's record purposes	Simultaneously with above					PS16

CONTRACT					JOB REFERENCE					
No.	Reg	Stage	Procedure	Description	Timing of action	Action Req		Initials	Date actioned	Sample document
						Yes	No			
16	*15(4)*		RECEIVE CONSTRUCTION PHASE HEALTH AND SAFETY PLAN	This must be forwarded by the principal contractor to the client or planning supervisor for assessment of its suitability. Whilst the planning supervisor 'shall be in a position to give adequate advice' to facilitate this process it remains the client's statutory duty to sanction its suitability.	BEFORE COMMENCEMENT OF CONSTRUCTION					PS17
17	*10*	Pre-construction	CONFIRM SUITABILITY OF CONSTRUCTION PHASE HEALTH AND SAFETY PLAN TO CLIENT[4]	It is imperative that commencement of construction does not start until a construction phase health and safety plan has been considered suitable. Once confirmed, client and principal contractor are to be notified	**BEFORE COMMENCEMENT OF CONSTRUCTION**					PS18
18		Construction	ENSURE HEALTH AND SAFETY MANAGEMENT IS ADDRESSED AS AN INTEGRAL AGENDA ITEM	Health and safety issues should be addressed at all progress meetings to monitor competence and enhance the culture. This agenda can be addressed by the contract administrator in the absence of the planning supervisor and copies forwarded accordingly	Progress meetings					PS19
19	*14(d)*	End of construction	**ENSURE PREPARATION OF HEALTH AND SAFETY FILE**	This is the health and safety maintenance/operational manual for the completed project and is usually collated and finalised by the planning supervisor with relevant information from • Client • Designer(s) • Principal contractor • Contractor(s)						PS21 (sample contents)
20	*14(f)*	End of construction	**ENSURE DELIVERY OF HEALTH AND SAFETY FILE TO CLIENT**	Since it is another key document, it should be delivered personally or sent by recorded delivery	Preferably in conjunction with project handover. DEFINITELY BEFORE FINAL ACCOUNT SETTLEMENT					PS20
21		End of construction	RECEIVE CLIENT CONFIRMATION THAT THE HEALTH AND SAFETY FILE HAS BEEN RECEIVED AND IS SATISFACTORY	This completes the planning supervisor's role and concludes his administrative obligations to the contract	Within 7 days of the delivery of the health and safety file					PS20 (confirmation slip)

[4] Once the suitability of the construction phase health and safety plan has been sanctioned, there is no obligation for the planning supervisor to subsequently appraise its development. It is acknowledged that existing contract administration procedures can address aspects relevant to the development of the construction phase health and safety plan, e.g. method statements and programme changes.

Section 5
THE DESIGNER

The designer

The designer is strategically placed to influence health and safety aspects of the project, not only from the creative opportunities offered by design but also from his or her position as professional adviser to the client. This is addressed within the Regulations through the duty to implement a risk assessment strategy and to inform the client of the client's duties in respect of the Regulations themselves (for in-house appointments this is not a requirement).

Furthermore, the influential position of the designer in health and safety management is acknowledged by the fact that the designer remains the one party whose duties are never exempt regardless of the status of the project. Thus, designers must always undertake an appropriate risk assessment strategy.

From the definition of design:

> 'design in relation to any structure includes drawing, design details, specification and bill of quantities (including specification of articles or substances) in relation to the structure',

and designer:

> 'designer means any person who carries on a trade, business or other undertaking in connection with which he prepares a design relating to a structure or part of a structure'

it is readily appreciated that many construction stakeholders can become readily involved in the design process.

Such stakeholders could include a client or contractor who alters or interferes with the design process through substitution of specification, etc. This could be further extended if recourse is taken of those involved in the design of components:

> 'Manufacturers supplying standard components which can be used in any project are not designers under CDM, although they may have duties under supply legislation. The person who selects the product is the designer under CDM and must take account of health and safety issues arising from its use. If a product is purpose-made for a project, the person who prepared the specification is a designer under CDM, and so is the manufacturer who develops the detailed design.'

This ensures that for those organisations involved in the fit-out of buildings the initial design role could be undertaken by the sales team who drive the process from the first point of contact with the client. Thus:

> 'Designers are those who have a trade or a business which involves them in:
>
> - preparing designs for construction work including variations – this includes preparing drawings, design details, specifications, bills of quantities and the specification of articles and substances, as well as the related analysis, calculations, and preparatory work; or
> - arranging for their employees or other people under their control to prepare designs relating to a structure or part of a structure.'

This latter point addresses the Construction (Design and Management) (Amendment) Regulations 2000, which relates to the Paul Werth case, where an anomaly highlighted by the Court of Appeal led to the amended regulations which now ensure that legal duties on designers to give consideration to health and safety as an integral part of design apply also to employees or other persons under designers' control, as originally intended.

Structure is subsequently defined as:

'(a) any building, steel or reinforced concrete structure (not being a building), railway line or siding, tramway line, dock, harbour, inland navigation, tunnel, shaft, bridge, viaduct, waterworks, reservoir, pipe or pipeline (whatever in either case, it contains or is intended to contain) cable, aqueduct sewage works, gasholder, road, airfield, sea defence, works, river works, drainage works, earthworks, lagoon, dam, wall, caisson, mast, tower, pylon, underground tank, earth retaining structure, or structure designed to preserve or alter any natural feature, and any other structure similar to the foregoing or,

(b) any formwork, falsework, scaffold or other structure designed or used to provide support or means of access during construction work, or

(c) any fixed plant in respect of work which is installation, commissioning, de-commissioning or dismantling and where any such work involves a risk of a person falling more than 2 metres.'

The duties to be discharged by the designer therefore apply to both permanent and temporary design and apply whenever the designer prepares a design 'for the purposes of construction work'.

Designers must:

- take reasonable steps to ensure that clients are aware of their duties under CDM before starting design work
- prepare designs with adequate regard to health and safety, and to the information supplied by the client
- provide adequate information in or with the design
- co-operate with the planning supervisor and with any other designers so that each of them can comply with their duties under the Regulations. This includes providing any information for the health and safety file.

Regardless of pedigree, designers must consider the need to design in a way which avoids risks to health and safety or reduces these risks as far as practicable so that projects they design can be constructed and maintained safely. Where risks cannot be avoided, suitable and sufficient information on them has to be provided to those who need that information.

Designers must be able to:

- Identify hazards inherent in their designs
- Identify the resultant risks during construction, maintenance or demolition
- Understand how to eliminate the hazards, or reduce the risks

Inevitably, this should be processed through the planning supervisor for inclusion in the health and safety plan and/or the health and safety file, but on some projects, there will not be a planning supervisor, because the project is a non-CDM status project. The designer's duties apply regardless and suitable information must be provided to those who require it so that the work can be undertaken safely. The annotated note on the sketch still remains a potent communication device.

Interestingly, the Approved Code of Practice does not advocate the legal necessity of keeping records of the design process, often referred to as the design risk assessment, but acknowledges that such records provide accountability in the event of demonstrating the exercise of reasonable professional judgement and compliance with CDM. This latter point receives support from CIRIA REPORT 166 which identifies the practical reasons for keeping records – in case of personnel changes, unexpected delays in the progress of the design, or in the case of future disputes.

Focused guidance for the designer has been provided in CIRIA REPORT 166, 'Work sector guidance for designers,' which identifies the need to provide information on the significant risks in the design, which cannot be avoided.

Whilst this simplifies the extent of the issues to be communicated to other parties it also directs attention to the subjective interpretation of the word significant. The prudent designer will note the detailed information provided in the Health and Safety Executive's annual Statistics Report, which identifies the three major categories of fatal statistics on construction sites. These are currently:

- falls from height
- contact with moving vehicles
- falling objects

Additionally it is to be noted that the largest category of occupational illness arises from musculo-skeletal disorders, which include manual handling aspects of the construction process. All these must form part of the consideration for health and safety in design.

Furthermore, the ACoP states that examples of significant hazards where designers always need to provide information include:

- Hazards that could cause multiple fatalities to the public, such as tunnelling, or the use of a crane close to a busy public place, major road or railway
- Temporary works, required to ensure stability during the construction, alteration or demolition of the whole or any part of the structure, e.g. bracing during construction of steel or concrete frame buildings
- Hazardous or flammable substances specified in the design, e.g. epoxy grouts, fungicidal paints, or those containing isocyanates
- Features of the design and sequences of assembly or disassembly that are crucial to safe working
- Specific problems and possible solutions, for example arrangements to enable the removal of a large item of plant from the basement of a building
- Structures that create particular access problems, such as domed access structures
- Heavy or awkward prefabricated elements likely to create risks in handling
- Areas needing access where normal methods of tying scaffolds may not be feasible, such as facades that have no opening windows and cannot be drilled.

The need for documentary support in respect of the design risk assessment strategy and potential future liabilities provides substantiation to the visibility of the process. This is normally manifested through the design risk assessment proforma, although a simpler tabulation of significant hazards and corresponding mitigation/control measures remains a satisfactory mechanism for simpler projects.

The fulfilment of the risk assessment strategy is initiated by the hazard assessment in respect of the project, translated into the risk appraisal phase and culminating in a design contribution to health and safety management focused on a hierarchy of response based on:

- Elimination
- Reduction
- Transfer

The risk assessment appraisal results in a risk assessment proforma which is effectively a synopsis of the strategy to date. The risk assessment proforma is a primary means of communication and should depict succinctly the design route and its health and safety contribution. Such risk assessments should be directed at significant and principal hazards specific to the project, and are not meant to become verbose and inappropriate from focusing on all foreseeable risks. The design risk assessment is based on the premise of the competent contractor and should not therefore dwell on those hazards which a competent contractor would ordinarily expect to manage. Throughout the design process, the designer must be able to answer the question 'What is your contribution to health and safety management?'

Further evidence of compliance with regulation 13 strategy is provided by:

- Design brief minutes
- Brainstorming sessions
- Team meetings
- Workshops

The effective discharge of duties under the CDM Regulations demands a proactive approach by the designer. The design challenge to contribute must be answered and early dialogue between the design team and the planning supervisor team can provide numerous benefits and opportunities for cooperation and contribution.

The designer can also function as the planning supervisor but extra vigilance is required to satisfactorily fulfil both roles and there is much to recommend the independence of the planning supervisor. Design and planning supervision multi-functional roles can remove impartiality and objectivity and prove difficult to fulfil.

Many inter-disciplinary projects carry numerous design and construction team members over lengthy phases of design and construction. It is essential for the co-ordination of health and safety management that important and significant issues raised during the design phase are duly closed down, with relevant action pursued to a satisfactory conclusion. Some of these issues could well impact on the workplace methodologies to be developed by the contractors in their contribution to the ongoing construction phase health and safety plan. Others could well lead to information forwarded for the collation of the health and safety file.

Thus on larger projects a portfolio of design risk assessments will be assembled. It is helpful if the filtering process associated with the significant and principal residual hazards abstracts the relevant collective information into a single risk register to facilitate ownership and closure of the important issues. Such a risk register is offered for consideration as sample document D2.

The design team must also provide relevant information for the purposes of the health and safety file. A focus is now provided towards information about key structural principles incorporated in the design of the structure as well as associated residual hazards. Additionally, it is useful if all design team members provide design philosophy statements to enable a better understanding of how health and safety management was approached by design team members. These could incorporate the strategies and methodologies that have been developed to account for:

- Work at height
- Fire prevention
- Maintenance
- Major component replacements, etc

5.1
DESIGNER FLOWCHART

Feasibility/design	**Design**	**Pre-tender**	**Construction**
RESPOND TO • COMPETENCE QUESTIONNAIRE • ALLOCATION OF ADEQUATE RESOURCES 1 *8(2)*	*Not undertaken for in-house appointments*		
INFORM CLIENT OF CDM RESPONSIBILITIES 2 *13(1)*	**This is an essential duty so that the lay client is duly informed of what the client's responsibilities are.** *Not required for in-house appointments*		
	ENSURE DESIGN HAS GIVEN ADEQUATE REGARD TO **• AVOIDANCE** **• PREVENTION** **• PRIORITISATION** 3 *13(2)(a)*	*Designers must contribute to health and safety through* • *Elimination* • *Reduction* • *Transfer* of associated hazards	
	ENSURE DESIGN INCLUDES ADEQUATE INFORMATION ON **• PROJECT** **• STRUCTURE** **• MATERIALS** 4 *13(2)(b)*	*This is based on reasonable enquiries being made*	
	COOPERATE WITH PLANNING SUPERVISOR AND OTHER DESIGNERS IN THE EXCHANGE OF INFORMATION 5 *13(2)(c)*	*A proactive response is required*	
	FORWARD RISK ASSESSMENT PROFORMAS[1] TO PLANNING SUPERVISOR 6	*This is a useful means of hazard communication. Alternatively, a list of residual hazards can be written out and forwarded separately*	UNDERTAKE RISK ASSESSMENTS OF RELEVANT VARIATION ORDERS AND ARCHITECT'S INSTRUCTIONS WHERE APPROPRIATE 7 *13(2)(a)*
			ENSURE RELEVANT DESIGN INFORMATION HAS BEEN FORWARDED TO PLANNING SUPERVISOR AND OTHER DESIGNERS[2] 8 *13(2)(c)*
			ENSURE 'AS BUILT' RECORD DRAWINGS ARE DRAFTED AND FORWARDED TO PLANNING SUPERVISOR 9 *13(2)(c)*

[1] Design and build procurement strategies require risk assessment methodologies to be fulfilled well into construction phases of such contracts.

[2] Where the project does not have CDM status, there will be no planning supervisor, and thus the designer has increased communication obligations in providing suitable and sufficient information to whoever requires it.

5.2 DESIGNER CHECKLIST

CONTRACT					JOB REFERENCE					
No.	Reg	Stage	Procedure	Description	Timing of action	Action Req		Initials	Date actioned	Sample document
						Yes	No			
1	*8(2)*	Feasibility/design	RESPOND TO • COMPETENCE QUESTIONNAIRE • ALLOCATION OF ADEQUATE RESOURCES	This is usually addressed through pre-qualification procedures when an 'Approved List' is drafted either annually or for the individual project. (NOT RELEVANT FOR IN-HOUSE APPOINTMENTS)	Prior to appointment as designer					PS2
2	*13(1)*	Feasibility/design	**INFORM CLIENT OF HIS DUTIES IN RESPECT OF THE CDM REGULATIONS**	**NOT RELEVANT FOR IN-HOUSE APPOINTMENTS**	Immediately on appointment					
3	*13(2)(a)* *13(2)(a)(i)* *13(2)(a)(ii)* *13(2)(a)(iii)*	Design	**ENSURE DESIGN HAS GIVEN ADEQUATE REGARD TO** • **AVOIDANCE OF FORESEEABLE RISKS** • **COMBATING RISKS AT SOURCE** • **PRIORITISATION OF MEASURES TO PROTECT ALL PERSONS**	This is achieved by the designer through his risk assessment strategy[1] based on the principles of prevention and protection. Documentary evidence should be available via risk assessment proformas that a CONTRIBUTION towards health and safety has been made based on considerations of • Elimination • Reduction • Transfer	Throughout design					D1 (example) Appendix A (completed)
4	*13(2)(b)*	Design	**ENSURE DESIGN HAS INCLUDED ADEQUATE INFORMATION ABOUT** • **PROJECT** • **STRUCTURE** • **MATERIALS**	Such information will be provided via appropriate statements on drawings and sketches and also through the risk assessment proformas. Innovative design particularly involving tensioned systems will require design method statements for downloading and decommissioning for inclusion in the health and safety file	Throughout all design phases to include the drafting of the specification					
5	*13(2)(c)*	Design	**COOPERATE WITH OTHER DESIGNERS AND PLANNING SUPERVISOR IN THE EXCHANGE OF INFORMATION RELEVANT TO HEALTH AND SAFETY MANAGEMENT**	This is essential to design communication throughout the project and will be coordinated by the planning supervisor[2]	Throughout all design phases					

[1] Risk assessment proformas represent vital documentation which will be appraised by the Health and Safety Executive in the event of an investigation.
[2] Design communication requirements need further consideration by the designer, particularly for a non-CDM status project where there is no planning supervisor. **Designer duties are never exempt.**

CONTRACT					JOB REFERENCE					
No.	Reg	Stage	Procedure	Description	Timing of action	Action Req		Initials	Date actioned	Sample document
						Yes	No			
6		Design/ pre-tender	FORWARD RISK ASSESSMENT PROFORMAS TO PLANNING SUPERVISOR	These forms enable the planning supervisor to confirm that a risk assessment strategy is in place and allows the residual risks to be transferred by the planning supervisor into the pre-tender health and safety plan and possibly into the health and safety file	Prior to the drafting of the pre-tender health and safety plan and throughout ongoing design where appropriate					
7	*13(2(a)*	Design	UNDERTAKE RISK ASSESSMENT OF RELEVANT VARIATION ORDERS OR ARCHITECT'S INSTRUCTIONS	The risk assessment strategy should also relate to health and safety issues arising from VOs and AIs or other design changes	Throughout the design process					
8	*13(2)(c)*	Design	**ENSURE RELEVANT INFORMATION HAS BEEN FORWARDED TO THE PLANNING SUPERVISOR AND OTHER DESIGNERS**	Suitable and sufficient information must be communicated for the purpose of health and safety management	Throughout the design process but completed by the time the health and safety file is collated					
9	*13(2)(c)*	Design/ construction	ENSURE 'AS BUILT' RECORD DRAWINGS ARE DRAFTED AND FORWARDED TO THE PLANNING SUPERVISOR PLUS OTHER RELEVANT INFORMATION, E.G. ● Downloading method statement ● Residual risks	These are for inclusion in the health and safety file and are factual representations for future reference in terms of maintenance and operation. Such drawings include schematic layouts of: ● drainage ● electrical works ● mechanical works, etc	Before project handover					
10		Post construction	Ensure all risk assessment proformas and other relevant documentary information are maintained for the full period of design liability							

Section 6
THE PRINCIPAL CONTRACTOR

The principal contractor

The principal contractor and his team can be seen as operating at the sharp end of construction and must ensure that productivity targets and deliverables associated with limited windows of opportunity do not compromise the delivery of health and safety management objectives.

Compliance with the CDM Regulations does not prohibit fast-track or accelerated programmes, but the corresponding level of health and safety management resourcing and competence must remain compatible.

The principal contractor's role is to manage and co-ordinate the construction phase. As with all the other functions it is based on competence and adequate resourcing, both of which must be established in conjunction with the procurement process.

Health and safety competence is usually assessed in conjunction with technical and financial competence and, in most situations, would be part of the pre-qualification process undertaken prior to tendering. The procedure adopted in drafting 'Approved Lists' ensures that this is done on a routine basis with re-appraisals undertaken at appropriate times. The competence issue is therefore effectively undertaken prior to invite to tender.

Adequacy of resourcing is project-specific and is directed at ensuring that the tendering contractor has allocated sufficient health and safety resources to undertake the project with SAFE SYSTEMS OF WORK and in a SUITABLE AND SUFFICIENT manner. These requirements are reflected in the tender price. The assessment of health and safety resourcing, however undertaken, should remove from consideration any contractor who has achieved a competitive edge simply by under-resourcing the health and safety management of the contract. Timing is critical and this requirement must be established prior to letting the contract.

The construction phase starts with the legal sanction provided by the client in respect of the suitability of the construction phase health and safety plan (regulation 10). This is a condition precedent to the start of the construction phase and it would be unwise for the enthusiastic principal contractor to commence construction in advance of sanction.

The main documentation related to the role of the principal contractor is the *construction phase health and safety plan*, which must be suitably developed before commencement of construction. The extent of this suitability is reflective of the nature of the project and will also be influenced by the interpretation of the planning supervisor. Preferably the planning supervisor should have provided some guidance on the initial content of the documentation with particular emphasis on the initial method statements required for that suitability. This is best conveyed in the *pre-tender health and safety plan.*

The prudent approach is to ensure that the construction phase health and safety plan is suitable and that the client has endorsed it to that effect in compliance with regulations 15(4) and 10.

The development of the construction phase health and safety plan compatible with the construction programme is a primary duty to be undertaken by the principal contractor. To expedite this development the principal contractor needs to provide guidance to the other contractors and sub-contractors about the presentation of their workplace method statements arising from their workplace risk assessments in advance of the associated work being carried out. This would form part of the site co-ordination process.

A major duty to be discharged concerns the terms of compliance with regulation 16(1)(c), which requires the principal contractor to:

> 'take reasonable steps to ensure that only authorised persons are allowed into any premises or part of premises where construction work is being carried out.'

This represents an essential but onerous task, particularly on extended sites where site boundaries can run for considerable distances. To discharge this duty the principal contractor must therefore take reasonable steps to ensure that the foreseeably uninvited person is prohibited from entry, since the law identifies construction sites as an inducement to children to trespass.

As with the clients' duty in respect of the start of construction, breach of this particular regulation confers a right of action in any civil proceedings.

The principal contractor should also note that whilst the construction phase health and safety plan is a dynamic document, continually evolving compatible with construction progress by the addition of method statements and information, he must continually endeavour to manage the construction process, effectively and proactively.

Proactive management remains critical to the health and safety success of every project with documentation an essential pre-requisite of management system support.

The appointment of contractors and sub-contractors must also be undertaken based on competence and allocation of adequate resourcing being established. The CDM Regulations also require the planning supervisor to be in a position to give adequate advice to both these areas should a contractor require such an input [Regulation 14(1)(c)].

Note also that the principal contractor in compliance with regulation 16(1)(d) must:

> 'ensure that the particulars [of the notification to the HSE] are displayed in a . . . position where they can be read by any person at work on construction work'.

This is usually by copy in the site office so that other parties can have access to the relevant information contained in the notification.

A key communication route for the control of information is the induction process, which is the acknowledged vehicle for providing the comprehensive range of information relevant to the site and the project, which the principal contractor needs to convey to other contractors, sub-contractors and the self-employed.

The principal contractor must also fulfil duties to other construction related legislation, particularly the Management of Health and Safety at Work Regulations 1999 and the Construction (Health, Safety and Welfare) Regulations 1996. Numerous other regulations usually apply.

The role of the principal contractor need not necessarily be a singular duty to be fulfilled under the CDM Regulations since many contractors also function as designers through their management of the design and build procurement process. Care should be exercised that all duties are effectively discharged.

Co-operation, communication and control remain key cornerstones for the successful health and safety management of the construction phase.

6.1
PRINCIPAL CONTRACTOR FLOWCHART

Pre-tender	**Tender**	**Pre-construction**	**Construction**
RESPOND TO COMPETENCE QUESTIONNAIRE 1 *8(3)*	*This may already have been undertaken via an 'Approved List' procedure* *FOR IN-HOUSE APPOINTMENTS, COMPETENCE MUST BE SATISFIED THROUGH IN-HOUSE PROCEDURES*		
RECEIVE PRE-TENDER HEALTH AND SAFETY PLAN FROM PLANNING SUPERVISOR 2 *15(1)*	*This should be received with tender documentation*		
	ENSURE TENDER SUBMISSION ACCOUNTS FOR ADEQUACY OF RELEVANT HEALTH AND SAFETY MANAGEMENT RESOURCES 3	*Health and safety management resources should be reflected in the tender price submitted*	
		RESPOND TO INTERVIEW SESSION FOR ALLOCATION OF ADEQUATE RESOURCES 4 *9(3)*	*Other methods other than interview can be adopted, namely:* • *Resource breakdown* • *Bill of quantities*
		RETURN MINUTES OF ALLOCATION OF ADEQUATE RESOURCE MEETING 5 *9(3)*	
		SIGN ADDITIONAL HSE NOTIFICATION (F10) 6 *7(4)*	*Usually done at pre-contract meeting. The principal contractor requires a copy for display on site in a prominent position*
		PREPARE CONSTRUCTION PHASE HEALTH AND SAFETY PLAN AND FORWARD TO CLIENT AND/OR PLANNING SUPERVISOR 7 *15(4)*	*This represents a framework for the health and safety management of the construction process and must continue to be developed throughout the construction phase*
		RECEIVE SANCTION OF CONSTRUCTION PHASE HEALTH AND SAFETY PLAN FROM CLIENT 8 *10*	**Commencement of construction is dependent upon the construction phase health and safety plan being deemed suitable by the client**
			ENSURE COOPERATION BETWEEN CONTRACTORS 9 *16(1)(a)*

Pre-tender	Tender	Pre-construction	Construction
			ENSURE EVERY CONTRACTOR AND EMPLOYEE COMPLIES WITH RULES 10 *16(1)(b)*
		Non-compliance with this regulation confers an immediate Civil Right of Action	**ENSURE ONLY AUTHORISED PERSONS ALLOWED ON PREMISES** 11 *16(1)(c)*
		Must be displayed in a prominent position on site where they can be read by any person at work	**DISPLAY ADDITIONAL HSE NOTIFICATION (F10)** 12 *16(1)(d)*
		Based on reasonable enquiries made of a contractor for inclusion in the health and safety file	**PROVIDE PLANNING SUPERVISOR WITH RELEVANT HEALTH AND SAFETY INFORMATION** 13 *16(1)(e)*
		This can be done via the construction phase health and safety plan or in the furnishing of relevant abstracts from it	**GIVE REASONABLE DIRECTIONS TO CONTRACTORS** 14 *16(2)(a)*
		This can be done via the construction phase health and safety plan or in the furnishing of relevant abstracts from it	**BRING SITE RULES TO ATTENTION OF PEOPLE AFFECTED BY THEM** 15 *16(3)*
		This can be done via the construction phase health and safety plan or in the furnishing of relevant abstracts from it NOTE: MUCH OF THIS INFORMATION CAN BE TRANSMITTED VIA THE SITE INDUCTION TALK – PRIOR TO STARTING WORK ON SITE	**PROVIDE EVERY CONTRACTOR WITH COMPREHENSIBLE INFORMATION ON HEALTH AND SAFETY** 16 *16(2)(b), 17(1)*
			ENSURE EVERY CONTRACTOR PROVIDES EMPLOYEES WITH INFORMATION ON ASSOCIATED CONTRACT RISKS TOGETHER WITH APPROPRIATE HEALTH AND SAFETY TRAINING 17 *17(2)*

Pre-tender	**Tender**	**Pre-construction**	**Construction**
		These are matters relevant to health and safety	**ENSURE EMPLOYEES AND SELF-EMPLOYED CAN DISCUSS AND OFFER ADVICE** 18 *18(a)*
			ENSURE ARRANGEMENTS EXIST FOR CO-ORDINATION OF VIEWS OF EMPLOYEES OR THEIR REPRESENTATIVE 19 *18(b)*
		The principal contractor must be proactive in the provision of such information	**PASS OVER RELEVANT INFORMATION TO PLANNING SUPERVISOR FOR COMPILATION OF HEALTH AND SAFETY FILE** 20 *16(1)(e)*
			NB. IT IS IMPERATIVE THAT THE PRINCIPAL CONTRACTOR CONTINUES TO DEVELOP THE CONSTRUCTION PHASE HEALTH AND SAFETY PLAN DURING THE CONSTRUCTION PHASE COMPATIBLE WITH THE CONSTRUCTION PROGRAMME 21

6.2
PRINCIPAL CONTRACTOR CHECKLIST

CONTRACT					**JOB REFERENCE**					
No.	**Reg**	**Stage**	**Procedure**	**Description**	**Timing of action**	**Action Req**		**Initials**	**Date actioned**	**Sample document**
						Yes	**No**			
1	*8(3)*	Pre-tender	RESPOND TO COMPETENCE QUESTIONNAIRE	No invite to tender should be extended to anyone whose competence has not been established	Before invitation to tender and often well in advance via the selection procedures associated with the 'Approved List'					PS7
2	*15(1)*	Pre-tender	RECEIVE PRE-TENDER HEALTH AND SAFETY PLAN FROM PLANNING SUPERVISOR	This informs the principal contractor of the significant health and safety issues specific to the project.	Simultaneously with tender documentation					PS6 (sample)
3		Tender	ENSURE TENDER SUBMISSION ACCOUNTS FOR ADEQUACY OF RELEVANT HEALTH AND SAFETY MANAGEMENT RESOURCES	All health and safety issues, particularly those outlined in the pre-tender health and safety plan, should be adequately resourced by the contractor and reflected in his tender price submission	During preparation of the tender					
4	*9(3)*	Post-tender/pre-letting	RESPOND TO ALLOCATION OF ADEQUATE RESOURCES	Allocation of adequacy of resourcing in terms of health and safety management will usually be established by the planning supervisor. This will possibly be undertaken via an interview session, minuted by the planning supervisor and sent to the principal contractor for signing and return[1] Other methods can be invoked for establishing resource adequacy	After tender scrutiny but **BEFORE THE LETTING OF THE CONTRACT. (For in-house appointments this could be addressed by a resource schedule associated with the estimate)**					PS10
5	*9(3)*	Post-tender/pre-letting	RETURN MINUTES OF ALLOCATION OF ADEQUATE RESOURCE MEETING	The minutes as received will be read, amended where appropriate, signed and returned by the principal contractor to the planning supervisor	After tender scrutiny prior to letting the contract					PS11 (sample interview), PS12
6	*7(4)*	Post-tender/pre-letting	**SIGN ADDITIONAL HSE NOTIFICATION (F10)**	The principal contractor must sign the relevant part of the F10 form as presented by the planning supervisor. This formally confirms the principal contractor's role and should be **displayed in a prominent position on site**	Simultaneously with letting of the contract					PS14

[1] Prior to signing the contract there exists the favoured contractor from amongst the tendering contractors. The principal contractor appointment is simultaneous with the signing of the contract.

CONTRACT					JOB REFERENCE					
No.	Reg	Stage	Procedure	Description	Timing of action	Action Req		Initials	Date actioned	Sample document
						Yes	No			
7	*15(4)*	Pre-construction	**PREPARE CONSTRUCTION PHASE HEALTH AND SAFETY PLAN AND FORWARD TO CLIENT AND/OR PLANNING SUPERVISOR FOR SANCTIONING[1]**	This is the development of the pre-construction health and safety plan and must be in a suitable form before permission to commence the works is given	**SANCTION OF ITS SUITABILITY BY THE CLIENT MUST BE RECEIVED PRIOR TO COMMENCEMENT OF CONSTRUCTION**					PC1 (sample contents)
	16(2)(b)			The plan must also include rules for the health and safety management of the construction work						
8	*10*	Pre-construction/ construction	**RECEIVE SANCTION TO START CONSTRUCTION**	This is the client's acknowledgement that the construction phase health and safety plan is suitable	**PRIOR TO COMMENCEMENT OF CONSTRUCTION**					PS18
9	*16(1)(a)*	Construction	**ENSURE COOPERATION BETWEEN CONTRACTORS**	The principal contractor must coordinate the health and safety management of all other contractors on site. This can be achieved through appropriate induction procedures and/or meetings	Throughout the construction phase					
10	*16(1)(b)*	Construction	**ENSURE EVERY CONTRACTOR COMPLIES WITH RULES**	The principal contractor must ensure that all contractors and employees comply with site rules as contained in the construction phase health and safety plan. THEY MUST THEREFORE RECEIVE RELEVANT ABSTRACTS OF THE CONSTRUCTION PHASE HEALTH AND SAFETY PLAN AS APPROPRIATE	Throughout the construction phase					
11	*16(1)(c)*	Construction	**ENSURE ONLY AUTHORISED PERSONS ALLOWED ON PREMISES**	The principal contractor must ensure that all unauthorised personnel are kept away from the premises where construction work is being undertaken. Due attention must therefore be given to suitable fencing, lockable gates and signs	Throughout the construction phase. NON-COMPLIANCE WITH THIS REGULATION CONFERS AN IMMEDIATE RIGHT OF CIVIL ACTION					

[1] The principal contractor is under an obligation to ensure the continued development of the construction phase health and safety plan compatible with construction progress, i.e. the construction phase health and safety plan remains a dynamic document.

CONTRACT					**JOB REFERENCE**					
No.	**Reg**	**Stage**	**Procedure**	**Description**	**Timing of action**	**Action Req**		**Initials**	**Date actioned**	**Sample document**
						Yes	**No**			
12	*16(1)(d)*	Construction	**DISPLAY ADDITIONAL HSE NOTIFICATION (F10)**	This must be displayed in a prominent position on site, to be read by any person involved in the construction work who wishes to see it	**ON COMMENCEMENT OF CONSTRUCTION WORKS AND CONTINUALLY THEREAFTER**					PS16
13	*16(1)(e)*	Construction	**PROVIDE PLANNING SUPERVISOR WITH RELEVANT HEALTH AND SAFETY INFORMATION**	The principal contractor acts here as a communication link between the other contractors and the planning supervisor, providing relevant information which he believes the planning supervisor does not have, and which could be included in the health and safety file	Throughout the construction phase					
				Such information would reflect the health and safety implications of future operations, maintenance and demolition and will include relevant • Design information • Hazardous data sheets • Maintenance schedules • Supplier addresses • Sub-contractor addresses • Test Certificates, etc.	Throughout construction BUT CONCLUDED BY PROJECT HANDOVER					
14	*16(2)(a)*	Construction	**GIVE REASONABLE DIRECTIONS TO CONTRACTORS**	The principal contractor coordinates health and safety management during construction and must therefore be proactive in providing reasonable directions as appropriate	Throughout the construction phase					
15	*16(3)*	Construction	**BRING SITE RULES TO ATTENTION OF PEOPLE AFFECTED BY THEM (SUCH RULES NEED TO BE IN WRITING)**	This again emphasises the need for the principal contractor to be proactive in coordinating health and safety. It is preferable to hand across at induction and get appropriate signature from each employee, sub-contractor, etc.	Throughout the construction phase					
16	*16(2)(b)* *17(1)*	Construction	**PROVIDE EVERY CONTRACTOR WITH COMPREHENSIBLE INFORMATION ON HEALTH AND SAFETY**	This is achieved by ensuring that all contractors are in receipt of the relevant parts of the construction phase health and safety plan before they start work. It can also be undertaken via an INDUCTION SESSION before the contractor comes on site	Prior to the contractor coming on site and throughout the construction phase					PC1

CONTRACT					JOB REFERENCE					
No.	Reg	Stage	Procedure	Description	Timing of action	Action Req		Initials	Date actioned	Sample document
						Yes	No			
17	*17(2)*	Construction	**ENSURE EVERY CONTRACTOR PROVIDES EMPLOYEES WITH INFORMATION ON ASSOCIATED CONTRACT RISKS TOGETHER WITH APPROPRIATE HEALTH AND SAFETY TRAINING**	This ensures that everyone coming into contact with the construction process has received the relevant information on the hazards associated with the project. Induction training, tool-box talks and related meetings are all suitable means of achieving this objective Contractors also contribute to the development of the construction phase health and safety plan by furnishing method statements appropriate to their procedures	Throughout the construction phase As the construction programme develops, interim induction sessions may be needed					PC1, PC2
18	*18(a)*	Construction	**ENSURE EMPLOYEES AND SELF-EMPLOYED CAN DISCUSS AND OFFER ADVICE**	This extends the communication link between employees, self-employed and the principal contractor. This can be easily undertaken with an appropriate agenda item at each progress meeting or via interim meetings	Throughout the construction phase					
19	*18(b)*	Construction	**ENSURE ARRANGEMENTS EXIST FOR COORDINATION OF VIEWS OF EMPLOYEES OR THEIR REPRESENTATIVES IN MATTERS OF HEALTH AND SAFETY**	This reinforces the requirement for total coordination of health and safety matters between everyone involved in the project and can be undertaken during progress meetings or via interim arrangements	Throughout the construction phase					
20	*16(1)(e)*	Construction	**PASS OVER RELEVANT INFORMATION TO PLANNING SUPERVISOR FOR HEALTH AND SAFETY FILE**	The majority of the information will become available towards the latter stages of the contract. Handover of information as it becomes available is recommended and provides some protection to contractor insolvency	As information becomes available					

In design and build procurement strategies, the principal contractor will also be a designer since the principal contractor is managing the design process. Ensure both functions are adequately fulfilled.

Section 7
THE CONTRACTOR

The contractor

Regulation 2(1) defines a contractor as:

> 'any person who carries on a trade, business or other undertaking (whether for profit or not) in connection with which he:
>
> (a) undertakes to or does carry out or manage construction work,
> (b) arranges for any person at work under his control (including where he is an employer, any employee of his) to carry out or manage construction work.'

The contractor must also demonstrate competence and adequacy of resourcing in a similar manner to the principal contractor. This is now manifesting itself through the sub-contractor 'Approved List' that many contractors are themselves establishing.

In compliance with the regulations the ACoP provides additional guidance in that all contractors must:

- Satisfy themselves that any contractors or designers they engage are competent and adequately resourced
- Co-operate with the principal contractor
- Provide information to the principal contractor about risks to others created by their work
- Comply with any reasonable directions from the principal contractor, and with any relevant rules in the health and safety plan
- Tell the principal contractor about accidents and dangerous occurrences
- Provide information for the health and safety file
- Ensure that projects for domestic clients are notified to the HSE in good time
- Provide information and training to their employees

In order for the contractor to be in a position to appreciate and manage the hazardous nature of the site environment an exchange of relevant information must occur between the contractor and the principal contractor. This is usually undertaken during the induction programme, which provides a control gateway for basic information to be exchanged so that the contractor can ensure controls are exercised and suitable and sufficient information is received for the development of safe systems of work. A similar arrangement between contractor and sub-contractor should suffice.

Effective communication is a pre-requisite for successful project management and the contractor/sub-contractor interface is as vital as the principal contractor/contractor interface in this respect. Information must cascade down, as essentially it moves from work-face upwards to the principal contractor.

As with other duty holders, the contractor must also discharge his/her duties in a proactive way. Particular attention is focused on the up-front provision of workplace risk assessments in compliance with regulation 3 of the Management of Health and Safety at Work Regulations 1999. These are critical to the principal contractor's role in ensuring the development of the construction phase health and safety plan and are often extended into a more specific response via corresponding method statements.

They are crucial to the health and safety management of the construction phase, providing an opportunity of planning the various activities and sub-activities compatible with the development of the construction programme.

The information interface between contractor and sub-contractor is vital to the promotion of health and safety management, and relevant abstracts both of the *pre-tender health and safety plan* and the *construction phase health and safety plan* provide such information.

It is important that both tendering contractors and sub-contractors receive relevant information at a time when they can account for it. This would be prior to tendering and at suitable stages afterwards as it arises. Contractors should not hesitate to request such information if it is not forthcoming since focused knowledge on the project environment and its associated hazards is critical to the management of the process.

The additional need for the notification of reportable accidents and diseases in compliance with the Reporting of Diseases and Dangerous Occurrences Regulations 1995 should also be noted in accordance with regulation 19(1)(e) of the CDM Regulations 1994.

Cooperation as well as communication remain critical throughout the health and safety management process with all parties having an explicit obligation to proactively supply relevant material particularly as required for the collation and completion of the *health and safety file*, such as certificates, health and safety maintenance information, hazard data sheets, etc.

7.1
CONTRACTOR FLOWCHART

Pre-tender	**Pre-construction**	**Construction**
RESPOND TO COMPETENCE QUESTIONNAIRE 1 *8(3)*	This may already have been undertaken via an 'APPROVED LIST' pre-qualification procedure	
RECEIVE RELEVANT ABSTRACTS OF PRE-TENDER HEALTH AND SAFETY PLAN 2 *15(1)*	These should be forthcoming from the principal contractor and are crucial to provide information on the hazards relevant to the project environment	
	ENSURE TENDER SUBMISSION ACCOUNTS FOR EACH RELEVANT HEALTH AND SAFETY MANAGEMENT RESOURCE 3 *9(3)*	
This should be received from the principal contractor prior to commencement on site	RECEIVE NOTIFICATION OF • NAME OF PLANNING SUPERVISOR • NAME OF PRINCIPAL CONTRACTOR • RELEVANT SECTIONS OF CONSTRUCTION PHASE HEALTH AND SAFETY PLAN 4 *19(2)*	**COOPERATE WITH PRINCIPAL CONTRACTOR** 5 *19(1)(a)*
	Based on 'so far as is reasonably practicable'	**PROMPTLY PROVIDE PRINCIPAL CONTRACTOR WITH INFORMATION RELEVANT TO HEALTH AND SAFETY** 6 *19(1)(b)*
	It is the principal contractor's role to ensure cooperation between all contractors for the benefits of health and safety management	**COMPLY WITH DIRECTIONS FROM PRINCIPAL CONTRACTOR** 7 *19(1)(c)*
	Each contractor should ensure that they have received relevant copies of site rules	**COMPLY WITH SITE RULES CONTAINED WITHIN CONSTRUCTION PHASE HEALTH AND SAFETY PLAN** 8 *19(1)(d)*
	Based on 'so far as is reasonably practicable'	**PROMPTLY PROVIDE PRINCIPAL CONTRACTOR WITH INFORMATION FOR NOTIFICATION OR REPORTING AS REQUIRED BY RIDDOR[1]** 9 *19(1)(e)*
	The contractor must be proactive in providing such information	**PROMPTLY PROVIDE INFORMATION TO PRINCIPAL CONTRACTOR FOR INCLUSION IN THE HEALTH AND SAFETY FILE** 10 *19(1)(f)*

[1] RIDDOR, Reporting of Injuries, Diseases and Dangerous Occurrences Regulations 1995.

Pre-tender	Pre-construction	Construction
	This vital information should be relayed with other tender information to allow tender submissions to account for adequate health and safety management	**PROVIDE EMPLOYEE WITH** • **NAME OF PLANNING SUPERVISOR** • **NAME OF PRINCIPAL CONTRACTOR** • **RELEVANT CONTENTS OF CONSTRUCTION PHASE HEALTH AND SAFETY PLAN** 11 *19(2), 19(4)* **PROVIDE SELF-EMPLOYED WITH** • **NAME OF PLANNING SUPERVISOR** • **NAME OF PRINCIPAL CONTRACTOR** • **RELEVANT CONTENTS OF CONSTRUCTION PHASE HEALTH AND SAFETY PLAN** 12 *19(3), 19(4)*

7.2
CONTRACTOR CHECKLIST

CONTRACT					**JOB REFERENCE**					
No.	**Reg**	**Stage**	**Procedure**	**Description**	**Timing of action**	**Action Req**		**Initials**	**Date actioned**	**Sample document**
						Yes	**No**			
1	*8(3)*	Pre-tender	RESPOND TO COMPETENCE QUESTIONNAIRE. (FOR IN-HOUSE APPOINTMENTS COMPETENCE WILL BE ADDRESSED BY STANDARD PROCEDURES AND WILL BE THE RESPONSIBILITY OF THE CLIENT)	This addresses the competence of the contractor in terms of health and safety management	Before invitation to tender and often well in advance via the selection procedures associated with the 'Approved List'					PS7
2	*15(1)*	Pre-tender	RECEIVE PRE-TENDER HEALTH AND SAFETY PLAN INFORMATION FROM THE PLANNING SUPERVISOR OR PRINCIPAL CONTRACTOR[1]	This informs the principal contractor of the significant health and safety issues specific to the project.	Simultaneously with tender documentation					PS6 (sample)
3	*9(3)*	Tender	ENSURE TENDER SUBMISSION ACCOUNTS FOR EACH RELEVANT HEALTH AND SAFETY MANAGEMENT RESOURCE AS REQUIRED	All health and safety issues, particularly those outlined in the pre-tender health and safety plan, should be adequately resourced by the contractor and reflected in his tender price submission	During preparation of the tender					
4	*19(2)*	Pre-construction	RECEIVE NOTIFICATION OF • NAME OF PLANNING SUPERVISOR • NAME OF PRINCIPAL CONTRACTOR • RELEVANT SECTIONS OF CONSTRUCTION PHASE HEALTH AND SAFETY PLAN	Most of this information would be contained on the F10 notice as displayed and also in the related information passed over by the principal contractor in the construction phase health and safety plan	Prior to starting construction work					
5	*19(1)(a)*	Construction	**COOPERATE WITH THE PRINCIPAL CONTRACTOR**	Every contractor shall be proactive in cooperating with the principal contractor on matters of health and safety management	Throughout the contract					
6	*19(1)(b)*	Construction	**PROMPTLY PROVIDE PRINCIPAL CONTRACTOR WITH INFORMATION RELEVANT TO HEALTH AND SAFETY**	Relevant health and safety management information shall be provided to the principal contractor. This includes relevant risk assessments and method statements appropriate to the work for inclusion into the principal contractor's construction phase health and safety plan	Throughout the contract					CONT 1

[1] For sub-contract tendering procedures relevant health and safety information will be made available by the main contractor (usually the principal contractor).

CONTRACT					JOB REFERENCE					
No.	Reg	Stage	Procedure	Description	Timing of action	Action Req		Initials	Date actioned	Sample document
						Yes	No			
7	*19(1)(c)*	Construction	**COMPLY WITH DIRECTIONS FROM PRINCIPAL CONTRACTOR**	The principal contractor coordinates health and safety management on site and all contractors must comply with reasonable instructions given	Throughout the contract					
8	*19(1)(d)*	Construction	**COMPLY WITH SITE RULES CONTAINED WITHIN THE CONSTRUCTION PHASE HEALTH AND SAFETY PLAN**	The relevant site rules must be passed across to every contractor by the principal contractor so that compliance can be achieved. Such site rules must be passed over before the contractor enters the site	Throughout the contract					
9	*19(1)(e)*	Construction	**PROMPTLY PROVIDE PRINCIPAL CONTRACTOR WITH INFORMATION FOR NOTIFICATION OR REPORTING AS REQUIRED BY RIDDOR**	All death, injury, conditions or dangerous occurrences requiring notification through RIDDOR must also be reported to the planning supervisor	Throughout the contract					CONT 2
10	*19(1)(f)*	Construction	**PROMPTLY PROVIDE INFORMATION TO PRINCIPAL CONTRACTOR FOR INCLUSION IN THE HEALTH AND SAFETY FILE**	This refers to information obtained by reasonable enquiries and which the principal contractor does not already have in his possession	Throughout the contract					
11	*19(2)*	Construction	**PROVIDE EMPLOYEE WITH** • **NAME OF PLANNING SUPERVISOR** • **NAME OF PRINCIPAL CONTRACTOR** • **RELEVANT CONTENTS OF CONSTRUCTION PHASE HEALTH AND SAFETY PLAN**	This ensures that the employee is adequately informed about the health and safety management aspects of the project	Prior to the employee commencing work on site					
12	*19(3)*	Construction	**PROVIDE SELF-EMPLOYED WITH** • **NAME OF PLANNING SUPERVISOR** • **NAME OF PRINCIPAL CONTRACTOR** • **RELEVANT CONTENTS OF CONSTRUCTION PHASE HEALTH AND SAFETY PLAN**	This ensures that the self-employed is adequately informed about the health and safety management aspects of the project	Preferably at the invite to tender stage and prior to commencing work on site					

CONTRACTOR CHECKLIST (Sheet 2 of 2)

Section 8
SAMPLE DOCUMENTS

GUIDE TO SAMPLE DOCUMENTS

Reference	Actioned by	Description	Required for/ from
Client			
CLIENT 1	Client	Client's agent competence questionnaire	Reference
CLIENT 2	Client	Planning supervisor competence and resource assessment questionnaire	Reference
CLIENT 3	Client	Service agreement: client/planning supervisor	PS
CLIENT 4	Client	Start of construction phase	Principal contractor
Planning supervisor			
PS1	Planning supervisor	F10 notification (initial) and accompanying letter	HSE/Client
PS2	Planning supervisor	Designer competence and resource assessment questionnaire and accompanying letter	Designer
PS3	Planning supervisor	Designer competence: request for additional information	Designer
PS4	Planning supervisor	Confirmation of designer competence and adequacy of resources	Client
PS5	Planning supervisor	Request for copy of risk assessment	Designer(s)
PS6	Planning supervisor	Sample copy of pre-tender health and safety plan	Reference
PS7	Planning supervisor	Contractor competence questionnaire and accompanying letter	Contractor(s)
PS8	Planning supervisor	Contractor competence: request for additional information	Contractor
PS9	Planning supervisor	Confirmation of contractor competence	Client
PS10	Planning supervisor	Request for addressing adequacy of resources	Principal contractor
PS11	Planning supervisor	Allocation of adequacy of resources: sample interview	Reference
PS12	Planning supervisor	Request for contractor's signature to minutes of adequacy of resources meeting	Contractor
PS13	Planning supervisor	Confirmation of allocation of adequate resources	Client
PS14	Planning supervisor	Request for principal contractor to complete relevant sections of F10 notification [original F10 to be amended (PS1)]	Principal contractor

Contd.

Contd.

Reference	Actioned by	Description	Required for/ from
Client continued			
PS15	Planning supervisor	Additional F10 notification	HSE/Client/ Principal contractor
PS16	Planning supervisor	F10 display	
PS17	Planning supervisor	Request for draft copy of construction phase health and safety plan	Client/Principal contractor
PS18	Planning supervisor	Sanction of suitability of construction phase health and safety plan	Principal contractor/ Client
PS19	Planning supervisor	Health and safety management agenda item: progress meeting	Client
PS20	Planning supervisor	Health and safety file accompanying letter and confirmation slip	Client/Principal contractor
PS21	Planning supervisor	Sample contents of health and safety file	Client
Designer			
D1	Designer	Example of risk assessment proforma (blank)	Reference
D2	Designer	Example of risk register	Reference
Principal contractor			
PC1	Planning supervisor	Construction phase health and safety plan: sample contents	Reference
PC2	Principal contractor	Hazard identification checklist and method statement proposal form	Reference
Contractor			
CONT 1	Contractor	Hazard identification checklist and method statement proposal form	Reference/ Principal contractor

CLIENT

CLIENT'S AGENT COMPETENCE QUESTIONNAIRE

Name of organisation: . . . .
Address: . . . .
. . . .
. . . . **Tel no.:** . . . .
Contact for further information: . . . .

(1) Briefly describe the types of work undertaken by your company/organisation:
. . . .

(2) Please enclose:
(a) A copy of your professional indemnity insurance
(b) A copy of your current general health and safety policy statement
(c) An outline of your management organisation structure with regard to allocation of duties, delegation of responsibilities, etc. in relation to the client's agent duties.

(3) Please forward details of all client's agent duties undertaken in the last 12 months, including type of contract, duration and tender value.

(4) Please provide details of health and safety resource facilities in support of the client's agent function:
e.g. Health and safety library
Management systems
Audit trail control mechanisms
Typical proformas.

(5) Please outline all relevant courses attended by your client's agent team members over the last 3 years.

(6) Please provide examples of your approach to the monitoring of competence in relation to the function of planning supervisor and designer.

(7) Please provide examples of your approach to the monitoring of competence in relation to the function of principal contractor and contractor.

(8) Has there been any Health and Safety Inspectorate action undertaken against any person and/or company involved in contracts where your organisation was functioning as client's agent under the CDM Regulations? Yes ☐ No ☐

If yes, please provide relevant information.

(9) Who in your organisation has executive responsibility for the management of health and safety?

(10) Please provide details of experience and qualification of the persons named in (10).

(11) Please provide details of the structure of your oganisation for this project in fulfilling the client's agent role. Who will provide advice on health and safety issues on this project?

Contd.

Contd.

(12) Please provide information on the scale of fees charged in undertaking the role of client's agent.

(13) Please provide details of experience, qualifications, membership of professional bodies, etc. and arrangements for continuing professional development of key staff employed on this project.

Signed: .... **Date:** ....

PLANNING SUPERVISOR COMPETENCE AND RESOURCE ASSESSMENT QUESTIONNAIRE

Name of organisation: . . . .
Address: . . . .
. . . .
. . . . **Tel no.:** . . . .
Contact for further information: . . . .

(1) Please enclose:
- (a) A copy of your professional indemnity insurance
- (b) A copy of your current general health and safety policy statement
- (c) An outline of your management organisation structure with regard to allocation of duties, delegation of responsibilities, etc. in relation to the planning supervision duties.

(2) Please provide details of the professional qualifications of your planning supervisor team.

(3) Please forward details of all planning supervision duties undertaken in the last 12 months, including type of contract, duration and tender value.

(4) Please provide details of health and safety resource facilities in support of the planning supervisor function:
e.g. Health and safety library
Management systems
Audit trail control mechanisms
Typical proformas.

(5) Please outline all relevant courses attended by your planning supervision team members over the last 3 years.

(6) Please outline your procedures for the coordination of health and safety management issues relevant to design.

(7) Please outline your procedures for ensuring a health and safety management contribution by the design team.

(8) Please provide examples of your approach to the monitoring of competence in relation to the function of principal contractor.

(9) Please provide an example of a pre-construction health and safety plan.

(10) Please provide details of any comments received from the Health and Safety Inspectorate in relation to your planning supervision service.

(11) Has there been any Health and Safety Executive action undertaken against any person and/or company involved in contracts where your organisation was functioning as planning supervisor? Yes ☐ No ☐

If yes, please provide relevant information.

Contd.

Contd.

(12) Please provide details of the structure of the planning supervision team for this project.

(13) Please provide information on the scale of fees charged in undertaking the role of planning supervisor.

Signed: . . . . **Date:** . . . .

SERVICE AGREEMENT: CLIENT/ PLANNING SUPERVISOR

Regs	Essential duties[1]	Optional duties[2]	Yes	No	Regs
7(1)	**ENSURE INITIAL NOTIFICATION TO HEALTH AND SAFETY EXECUTIVE F10 REV (A)**	Complete and forward initial notification to Health and Safety Executive			*7(3)*
14(c)(i) *14(c)(ii)*	**ENSURE AVAILABILITY TO GIVE ADVICE ON THE COMPETENCE AND RESOURCING OF DESIGNER(S)**	Establish competence of designer(s) Establish adequacy of resourcing of designer(s)			*8(2)* 9(2)
14(a)(i)	**ENSURE DESIGNER HAS UNDERTAKEN AN APPROPRIATE RISK ASSESSMENT APPROACH**	Develop and formulate a suitable risk assessment strategy			*14(a)(i)*
14(a)(ii)	**ENSURE DESIGNER INCLUDES ADEQUATE INFORMATION**	Collate and coordinate information between designers			*14(a)(ii)*
14(b)	**ENSURE COOPERATION BETWEEN DESIGNERS FOR COMPLIANCE WITH REGn 13**	Establish suitable management systems between designers for information exchange and cooperation			*14(b)*
14(c)(i) *14(c)(ii)*	**BE IN A POSITION TO GIVE ADEQUATE ADVICE ON COMPETENCE AND RESOURCE ISSUES IN RESPECT OF CONTRACTOR**	Establish competence of contractor(s) Establish adequacy of resourcing of contractor(s)			*8(3)* *9(3)*

Contd.

Contd.

Regs	Essential duties[1]	Optional duties[2]	Yes	No	Regs
15(1)	**ENSURE PRE-TENDER HEALTH AND SAFETY PLAN IS PREPARED**	Prepare pre-tender health and safety plan to accompany tender documentation			*15(1)*
7(i)	**ENSURE ADDITIONAL NOTIFICATION TO HSE ON APPOINTMENT OF PRINCIPAL CONTRACTOR**	Complete and forward additional notification to Health and Safety Executive			*7(4)*
14(c)(ii)	**ENSURE AVAILABILITY TO GIVE ADVICE ON THE SUITABILITY OF THE CONSTRUCTION PHASE HEALTH AND SAFETY PLAN**	Appraise, comment and confirm suitability of construction phase health and safety plan prior to commencement of construction			*10*
		Monitor competence of designer(s) and/or contractor(s) throughout project			
14(d)	**ENSURE HEALTH AND SAFETY FILE IS PREPARED**	Receive, collate and prepare the health and safety file			*14(d), 14(e)*
14(e)	**REVIEW, AMEND OR ADD TO THE HEALTH AND SAFETY FILE PRIOR TO DELIVERY TO CLIENT**				

Contd.

Contd.

Regs	Essential duties[1]	Optional duties[2]	Yes	No	Regs
14(f)	**ENSURE DELIVERY OF HEALTH AND SAFETY FILE TO CLIENT**	Deliver health and safety file to client			*14(l)*
		Undertake post-contract audit Prepare and present associated report			

Notes:

(1) Essential duties are those duties which are OBLIGATORY in order for the planning supervisor to fulfil his/her statutory obligations in compliance with the CDM Regulations.

(2) Optional duties reflect the additional service that can be provided by the planning supervisor in order to provide an effective response to the CDM Regulations.

Principal Contractor
[Address]

Ref:

[Date]

Dear Sir

Construction (Design and Management) Regulations 1994: Start of construction phase-regulation 10
Project: . . . .

Further to your provision of the construction phase health and safety plan, I am pleased to confirm compliance with regulation 15(4) of the Construction (Design and Management) Regulations 1994 and am therefore able to sanction the start of the construction phase in respect of this project.

Yours faithfully

Client

PLANNING SUPERVISOR

F10 NOTIFICATION (INITIAL) AND ACCOMPANYING LETTER

Health and Safety Executive Ref:
[Address]

[Date]

Dear Sir

Construction (Design and Management) Regulations 1994: F10 initial notification
Project: . . . .

Please find enclosed form F10 for initial notification of the above project in compliance with regulation 7(3) of the Construction (Design and Management) Regulations 1994.

Yours faithfully

Planning Supervisor
for and on behalf of

Enc. F10 (initial)

cc. [Client]

NOTIFICATION OF PROJECT

Client	V	PV	NV	Planning supervisor	V	PV	NV
Focus serial number				Principal contractor	V	PV	NV

(1) Is this the initial notification of this project or are you providing additional information that was not previously available?

Initial notification ☐ Additional notification ☐

(2) **Client:** name, full address, postcode and telephone number (*if more than one client, please attach details on separate sheet*).
Name: Tel no.:
Address:
Postcode:

(3) **Planning supervisor:** name, full address, postcode and telephone number.
Name: Tel no.:
Address:
Postcode:

(4) **Principal contractor** (*or contractor when project is for a domestic client*): name, full address, postcode and telephone number.
Name: Tel no.:
Address:
Postcode:

(5) **Address of site:** (*where construction work is to be carried out*).
Address:
Postcode:

(6) **Local authority:** name of the unitary authority or island council within whose district the operations are to be carried out.

(7) Please give your estimates on the following: please indicate if these estimates are original.

☐ Original ☐ Revised (*tick relevant box*)

(a) The planned date for the commencement of the construction ☐
(b) How long the construction work is expected to take (in weeks) ☐
(c) The maximum number of people carrying out construction work on the site at any one time ☐
(d) The number of contractors expected to work on site ☐

V, visit; PV, possible visit; NV, no visit. (This is for HSE administration only.)

Contd.

Contd.

(8) **Construction work:** please give brief details of the type of construction work that will be carried out.

(9) **Contractors:** name, full address and postcode of those who have been chosen to work on the project (if required continue on a separate sheet). (Note: this information is only required when it is known at the time notification is first made to the Health and Safety Executive. An update is not required.)

(10) **Declaration of planning supervisor**
I hereby declare that .... has been appointed as planning supervisor for this project.
Signed by or on behalf of the organisation ....
(print name) ....
Date: ....

(11) **Declaration of principal contractor**
I hereby declare that .... (name of principal contractor) has been appointed as principal contractor for the project.
Signed by or on behalf of the organisation ....
(print name) ....
Date: ....

DESIGNER COMPETENCE AND RESOURCE ASSESSMENT QUESTIONNAIRE AND ACCOMPANYING LETTER

[Designer] Ref:
[Address]

[Date]

Dear Sir

Construction (Design and Management) Regulations 1994: Designer competence and adequacy of resourcing – regulations 8(2) and 9(2)
Project: . . . .

Further to your involvement in the above project, it is a statutory requirement that the competence and adequacy of resourcing of the designer are established in compliance with regulations 8(2) and 9(2) of the Construction (Design and Management) Regulations 1994.

I would therefore ask that you complete, sign and provide the requisite information, as requested in the attached questionnaire, as soon as possible.

Please contact me if I can be of further assistance.

Yours faithfully

Planning Supervisor
for and on behalf of

Enc. Designer competence and resource assessment questionnaire

DESIGNER COMPETENCE AND RESOURCE ASSESSMENT QUESTIONNAIRE

Name of organisation: . . . .
Address: . . . .
. . . . **Tel no.:** . . . .
Contact for further information: . . . .

It is not essential that you are able to answer all the following questions, but the more questions that are answered the more effective the assessment process.

General policy

(1) Please enclose:
  (a) A copy of your current general health and safety policy statement
  (b) An outline of your management organisation structure with regard to allocation of duties, delegation of responsibilities, etc. in relation to health and safety.

(2) Is each member of your organisation aware of his/her responsibilities under the Construction (Design and Management) Regulations 1994? Yes ☐ No ☐

(3) Please provide the name of any group, body, organisation or similar which you are a member and which promotes and has an involvement in health and safety matters.

(4) Please provide the names of all key personnel within your organisation who have attended a health and safety course within the last 3 years.

(5) Are you aware of your responsibility to notify the client of his/her responsibilities under the Construction (Design and Management) Regulations 1994 in compliance with regulation 13(1)? Yes ☐ No ☐

(6) Do you currently undertake a risk assessment approach as part of your duties under the Management of Health and Safety at Work Regulations 1999? Yes ☐ No ☐

(7) Please enclose any standard forms used for such risk assessments.

(8) Are you aware of the need for the designer to contribute to health and safety management through a proactive response based on: Yes ☐ No ☐
  Elimination
  Reduction
  Transfer

Organisation

(9) Please provide details of health and safety training undertaken by members within your organisation.

(10) Please outline your organisational procedures for disseminating up-to-date developments on health and safety issues.

(11) Please provide details of the organisational structure relevant to this project.

Contd.

Contd.

Planning and monitoring

(12) Please outline your organisational structure for the coordination of health and safety information between the design team and the planning supervisor.

(13) Please outline the arrangements for the coordination of information between the lead design member and other members of the interdisciplinary team.

(14) Does your organisation currently undertake a post-contract review of health and safety matters? Yes ☐ No ☐

(15) Please outline the coordination links for the dissemination of health and safety information on this project.

Resources

(16) Please provide details of health and safety resources available for reference and support in the health and safety management of this contract.

(17) Please outline any specialist resources which are utilised by your organisation in an advisory capacity on health and safety matters.

(18) Does your organisation have a comprehensive health and safety library in support of its health and safety management design function? Yes ☐ No ☐

Perspective

(19) Are you aware of your obligation to cooperate with the planning supervisor and any other designer in relation to the health and safety management of the project? Yes ☐ No ☐

(20) Do you consider your organisation to be competent and adequately resourced to fulful its obligations under the Construction (Design and Management) Regulations 1994? Yes ☐ No ☐

Signed: .... **Date:** ....
Designation: ....

DESIGNER COMPETENCE: REQUEST FOR ADDITIONAL INFORMATION

[Designer]
[Address]

Ref:

[Date]

Dear Sir

Construction (Design and Management) Regulations 1994: Designer competence and adequacy of resourcing – regulations 8(2) and 9(2)
Project: . . . .

Further to the information provided in response to the designer competence and resource assessment questionnaire, I am unable to confirm satisfaction in compliance with regulations 8(2) and 9(2) of the Construction (Design and Management) Regulations 1994.

Please provide additional information in relation to:

General policy
Organisation
Planning and monitoring
Resources
Perspective

Further appraisal will be undertaken once the requisite information is forthcoming.

Yours faithfully

Planning Supervisor
for and on behalf of

CONFIRMATION OF DESIGNER COMPETENCE AND ADEQUACY OF RESOURCES

[Client]
[Address]

Ref:

[Date]

Dear Sir

Construction (Design and Management) Regulations 1994: Designer competence and adequacy of resourcing – regulations 8(2) and 9(2)
Project: . . . .

Further to the information provided by . . . . in relation to the designer competence questionnaire, I am reasonably satisfied that the designer has the competence and adequate resources to prepare the design in compliance with regulations 8(2) and 9(2) of the Construction (Design and Management) Regulations 1994.

Yours faithfully

Planning Supervisor
for and on behalf of

cc. [Designer(s)]

REQUEST FOR COPY OF RISK ASSESSMENT

[Designer]
[Address]

Ref:

[Date]

Dear Sir

Construction (Design and Management) Regulations 1994: Risk assessment – regulation 13(2)
Project: . . . .

Further to your design function on the above project, please provide a copy of your risk assessment(s) as undertaken in compliance with regulation 13(2) of the Construction (Design and Management) Regulations 1994.

Yours faithfully

Planning Supervisor
for and on behalf of

SAMPLE COPY OF PRE-TENDER HEALTH AND SAFETY PLAN

CLIENT:

[Name]

CONTRACT:

e.g. Installation and commissioning of ready-mixed concrete batching plant

LOCATION:

[Quarry name and address]

CONTENTS[1]

[1] The format complies with that set out in Appendix 3 of the Approved Code of Practice.

The contractor is informed that the health and safety plan is drafted in compliance with the requirements of the Construction (Design and Management) Regulations 1994.

It identifies the significant health and safety management issues within the contract such that the contractor can adequately resource these issues in addition to the health and safety management aspects of the contract which a competent contractor would ordinarily resource under the Health and Safety at Work etc. Act 1974 and associated legislation.

(1) DESCRIPTION OF PROJECT

(a)	Location:	[Quarry name and address]
	Project description:	Installation and commissioning of a prefabricated ready-mixed concrete batching plant, including associated service provision.
	Programme details:	Commencement: 8 July 2002 Duration: 8 weeks
(b)	Client: Designer: Planning supervisor:	[Name] [Name] [Name]
(c)	Existing records	The general arrangement of the quarry and working faces is illustrated on drawing numbers (as included in tender documentation): L71 dc-2-89 [date] – [name of] Quarry Sketch .... Plant elevations Mechweld Engineering drawing: CHEP 1297/B ... Sheet 1 } Setting out information Sheet 2 } and elevations A geological report reference no. [give no. and date] is available for information in the quarry manager's office

2. CLIENT'S CONSIDERATIONS AND MANAGEMENT REQUIREMENTS

(a) Structure and organisation:

- An organigram of site management duties and responsibilities is attached.
- The Quarry Manager is the ultimate daily authority within the curtilage of the site.
- In his absence the Deputy Quarry Manager will function in that capacity.
- Such absence will be notified to the Principal Contractor in writing.

(b) Safety goals:

- The site management seeks to consolidate its stated aim of achieving, as a minimum, an ALL INCIDENT FREQUENCY RATE of better than 4.0/100,000 man-hours.
- All contractors on site will be expected to achieve the same.

(c) Permits and authorisation:

- All contractor's staff and operatives will be identified via a daily inventory to the Quarry Manager.

Contd.

Contd.

- **THE CONSTRUCTION PHASE HEALTH AND SAFETY PLAN DEVELOPED FROM THE PRE-TENDER HEALTH AND SAFETY PLAN MUST BE SUBMITTED TO THE PLANNING SUPERVISOR NOT LESS THAN ONE WEEK BEFORE THE PROPOSED DATE FOR START OF CONSTRUCTION WORK.**
- **NO CONSTRUCTION WORK IS TO COMMENCE UNTIL CONFIRMATION HAS BEEN RECEIVED IN WRITING ON BEHALF OF THE EMPLOYER THAT THE CONSTRUCTION PHASE HEALTH AND SAFETY PLAN IS DEEMED SUITABLE IN COMPLIANCE WITH REGULATION 15(4).**

- The construction phase health and safety plan to be submitted must include (as appropriate) the following:

1. Description of the project
   - Project description and programme details
   - Details of client, designers, planning supervisor and other consultants
   - Extent and location of existing records and plans
2. Communication and management of the work
   - Management structure and responsibilities
   - Health and safety goals for the project and arrangements for monitoring and review of health and safety performance
   - Arrangements for:
     - regular liaison between parties on site
     - consultation with the workforce
     - the exchange of design information between the client, designers, planning supervisor and contractors on site
   - Handling design changes during the project
   - The selection and control of contractors
   - The exchange of health and safety information between contractors
   - Security, site induction and on-site training
   - Welfare facilities and first aid
   - The reporting and investigation of accidents and incidents including near misses
   - The production and approval of risk assessments and method statements

   SITE RULES

   FIRE AND EMERGENCY PROCEDURES
3. Arrangements for controlling significant site risks

   SAFETY RISKS:
   - Services including temporary electrical installations
   - Preventing falls
   - Work with or near fragile materials
   - Control of lifting operations

Contd.

Contd.

	• Dealing with services (water, electricity and gas) • The maintenance of plant and equipment • Poor ground conditions • Traffic routes and segregation of vehicles and pedestrians • Storage of hazardous materials • Dealing with existing unstable structures • Accommodation adjacent land use • Other significant safety risks **HEALTH RISKS:** • Removal of asbestos • Dealing with contaminated land • Manual handling • Use of hazardous substances • Reducing noise and vibrations • Other significant health risks
4. The health and safety file	LAYOUT AND FORMAT ARRANGEMENTS FOR THE COLLECTION AND GATHERING OF INFORMATION STORAGE OF INFORMATION

(d) Emergency procedures:

- In the event of the emergency siren being sounded all individuals will vacate the premises and assemble at the north end of the main car park.

(e) Site rules:

- The existing environment is controlled by the Quarry Manager and is subject to both 'Generic Manager's Rules', and 'Manager's Site-Specific Rules'.
- **THE QUARRY MANAGER'S DECISION IS BINDING IN RELATION TO PRIORITISING OF PROCEDURES**
- Working hours shall be between 0800 and 1700 hours, Monday to Friday.
- Pollution of water-courses is prohibited.
- The contractor shall not permit employees to use radios or other portable radio equipment within the curtilage of the quarry.
- All deliveries are to be supervised by the contractor's banksman.
- The contractor is to take all reasonable precautions to prevent nuisance from smoke, rubbish, vermin, dust, etc.
- Smoking will not be permitted on the site except in mess rooms, which must be carefully controlled and inspected to guard against the risk of fire.
- No discharge of any kind will be allowed into existing water-courses.
- No fires to be lit on site.
- All waste will be disposed of off site.

Contd.

Contd.

(f) Activities on or adjacent to the site:

- **THE CONTRACTOR SHOULD NOTE THAT THE WORKS ARE TO BE CARRIED OUT IN A WORKING ENVIRONMENT.**
- Temporary site accommodation is limited to the area highlighted on sketch number: SX/LQ/1000.

(g) Liaison arrangements between parties:

- **THE CONTRACTOR SHALL LIAISE CLOSELY WITH THE QUARRY MANAGER IN RELATION TO THE PROGRAMME OF CONSTRUCTION WORKS AND NORMAL QUARRYING OPERATIONS.**
- **BLASTING OPERATIONS OCCUR DAILY AND MUST BE ACCOUNTED FOR THROUGH LIAISON WITH THE QUARRY MANAGER.**
- The Quarry Manager is [name], telephone number [provide number].
- The Planning Supervisor is [name], telephone number [provide number].

(h) Security arrangements:

- All vehicles must report to the weigh-bridge office.
- All vehicles must be parked within the confines of the designated area within the quarry.
- All contractor's operatives and staff shall display their contractor's identification pass at all times.

3. Environmental restrictions and existing on-site risks

(a) Boundaries and access:

- The site entrance is off the [give site location] road and via a sloping single track metalled road with passing places.
- A 17.5 tonne weight limit is imposed on the [give road number] at a point 5 miles north of the quarry entrance towards [location].
- The site boundary is well defined and surrounded on three sides by the River [name], which is tidal in this area.
- A deep-water drainage sump is sited close to the entrance to the batching plant location.

Adjacent land uses:

- The adjacent land use is a mixture of sheep-farming and forestry.

Existing storage of hazardous materials:

- A bituminous coating plant is operational within the quarry and located close to the position of the proposed batching plant.

Contd.

Contd.

Location of existing services:
- **ALL SERVICES ARE TO BE TREATED AS LIVE UNTIL PROVEN OTHERWISE.**
- **ALL SERVICES ARE TO BE ISOLATED PRIOR TO ASSOCIATED WORK.**
- All service positions are to be identified prior to commencement of works.

Ground conditions:
- The ground conditions comprise a pennant sandstone rock escarpment arising from quarrying operations.
- Further information can be obtained from the geological report reference no. [give no. and date] available for information in the quarry manager's office.

Existing structures:
- The location of the batching plant is adjacent to a 30 metre vertical rock face.
- The rock face is subject to weathering and rock falls are a potential hazard.

(b) Asbestos:
- Amosite asbestos identified in the main offices has been totally removed as part of the company's asbestos removal initiative undertaken in April 1998.

Existing storage of hazardous materials:
- Bituminous products storage tanks and a chemical store are located in close proximity to the proposed batching plant.
- The contractor's attention is drawn to the hazards associated with:
  - bituminous materials
  - chemical storage
  - contaminated groundwater
  - paint systems (chromates)

Contaminated land:

Leakage has occurred from the bituminous products storage tanks and hydrocarbon contamination exists (refer to report no. [give no.]).

Existing structures hazardous materials:
- Refer to comment under asbestos section above.

Health risks arising from client's activities:
- **REGULAR BLASTING OPERATIONS OCCUR AS PART OF THE AGGREGATE PRODUCTION PROCESS.**

Contd.

Contd.

4. **Significant design and construction hazards**

(a) **Design assumptions and control measures:**

- Temporary bracing must be provided until minimum concrete strengths are achieved in the foundations as identified on drawing no. CHEP 1297/B.

(b) **Design liaison:**

- Design alterations to be referred back to [give name and telephone number] at the design office.

(c) **Significant and principal risks:**

These include:

- Blasting operations
- Rock falls from rock face
- Deep water associated with drainage sumps
- Moving vehicles
- Work at height
- Stored materials
- Contaminated groundwater

(d) **Materials requiring particular precautions:**

- Epoxy resin hardeners in anchor bolts
- Protective painting systems to steelwork

ALL MATERIALS TO BE STORED AND USED IN ACCORDANCE WITH MANUFACTURERS' INSTRUCTIONS AND CORRESPONDING COSHH ASSESSMENTS.

5. **The health and safety file**

(a) **Format and content:**

It is a requirement of the Regulations that the principal contractor implements an effective management system by which the requisite information is provided for inclusion in the health and safety file, which should include as appropriate:

- A brief description of the work carried out
- Residual hazards and how they have been dealt with (for example surveys or other information concerning asbestos, contaminated land, water-bearing strata, buried services)
- Key structural principles incorporated in the design of the structure (e.g. bracing, sources of substantial stored energy – including pre-or post-tensioned members) and safe working loads for floors and roofs, particularly where these may preclude placing scaffolding or heavy machinery there
- Any hazards associated with the materials used (for example hazardous substances, lead paint, special coatings which should not be burned off)

Contd.

Contd.

- Information regarding the removal or dismantling of installed plant and equipment (for example lifting arrangements)
- Health and safety information about equipment provided for cleaning or maintaining the structure
- The nature, location and markings of significant services, including fire fighting services
- Information and as-built drawings of the structure, its plant and equipment (e.g. the means of safe access to and from service voids, fire doors and compartmentation)

CONTRACTOR COMPETENCE QUESTIONNAIRE AND ACCOMPANYING LETTER

[Contractor]
[Address]

Ref:

[Date]

Dear Sir

Construction (Design and Management) Regulations 1994: Contractor competence – regulation 8(3)
Project: . . . .

Further to your involvement in the above project, it is a statutory requirement that the competence of the contractor is established in compliance with regulation 8(3) of the Construction (Design and Management) Regulations 1994.

I would therefore ask that you complete, sign and provide the requisite information as requested in the attached questionnaire, as soon as possible.

Please contact me if I can be of further assistance.

Yours faithfully

Planning Supervisor
for and on behalf of

Enc. Contractor competence assessment questionnaire

CONTRACTOR COMPETENCE ASSESSMENT QUESTIONNAIRE

Name of organisation: . . . .
Address: . . . .
. . . .
. . . . **Tel no.:** . . . .
Contact for further information: . . . .

It is not essential that you are able to answer all the following questions, but the more questions that are answered the more effective the assessment process.

General policy

(1) Please enclose:
   (a) A copy of your current general health and safety policy statement ☐
   (b) An outline of your management organisational structure with regard to allocation of duties, delegation of responsibilities, etc. in relation to health and safety. ☐

(2) Is each member of your organisation aware of his/her responsibilities under the Construction (Design and Management) Regulations 1994? Yes ☐ No ☐

(3) Please provide the names of all key personnel within your organisation who have attended a health and safety course within the last 3 years. ☐

(4) Please provide information on all prohibition, improvement or other enforcement notices issued against your company within the past 5 years. ☐

(5) Please provide information on any awards received for accident prevention within the last 5 years. ☐

(6) Please provide summaries of statistical information in relation to company accidents, injuries and dangerous occurrences over the last 3 years. ☐

(7) Please provide details of any prosecution(s) undertaken against your company or individuals employed by your company for breaches of health and safety legislation within the past 5 years. ☐

Organisation

(8) Please provide details of health and safety training undertaken by members within your organisation. ☐

(9) Please outline your organisational procedures for disseminating information on up-to-date developments in health and safety issues. ☐

(10) Please provide the names and qualifications of internal safety advisers and/or external safety consultants used by your organisation. ☐

(11) Are there safety representatives appointed within your workforce? Yes ☐ No ☐

(12) Do you have a safety committee for consultation purposes? Yes ☐ No ☐

Contd.

Contd.

(13) Do you currently undertake risk assessment appraisals as part of your duties under the Management of Health and Safety at Work Regulations 1999 and the Construction (Health, Safety and Welfare) Regulations 1996? Yes ☐ No ☐

(14) Please enclose any standard forms used for undertaking such risk assessments. ☐

Planning and monitoring

(15) Does your organisation currently undertake a post-contract review of health and safety management? Yes ☐ No ☐

(16) Please outline your organisational structure for the coordination of health and safety information between:
(a) Principal contractor and the planning supervisor ☐
(b) Principal contractor and sub-contractor ☐

Resources

(17) Please outline any specialist resources which are utilised by your organisation in an advisory capacity on health and safety matters. ☐

(18) Do you undertake competence assessments of all your sub-contractors prior to appointment? Yes ☐ No ☐

(19) How do you monitor the competence and effective resourcing of your sub-contractors in health and safety matters? ☐

Perspective

(20) Are you aware of the need for the principal contractor to provide suitable information on risks to health and safety for employees and associated sub-contractors? Yes ☐ No ☐

(21) Do you consider your organisation to be competent and adequately resourced to fulfil its obligations under the Construction (Design and Management) Regulations 1994? Yes ☐ No ☐

(22) Are you aware that the construction phase health and safety plan must be deemed suitable before the commencement of construction? Yes ☐ No ☐

Signed: . . . . **Date:** . . . .
Designation: . . . .

CONTRACTOR COMPETENCE: REQUEST FOR ADDITIONAL INFORMATION

[Contractor] Ref:
[Address]

[Date]

Dear Sir

Construction (Design and Management) Regulations 1994: Contractor competence – regulation 8(3)
Project: . . . .

Further to the information provided in response to the contractor competence assessment questionnaire, I am unable to confirm satisfaction in compliance with regulation 8(3) of the Construction (Design and Management) Regulations 1994.

Please provide additional information in relation to:

General policy

Organisation

Planning and monitoring

Resources

Perspective

Further appraisal will be undertaken once the requisite information is forthcoming.

Yours faithfully

Planning Supervisor
for and on behalf of

CONFIRMATION OF CONTRACTOR COMPETENCE

[Client] Ref:
[Address]

[Date]

Dear Sir

Construction (Design and Management) Regulations 1994: Contractor competence – regulation 8(3)
Project: . . . .

Further to the information provided by . . . . in relation to the contractor competence assessment questionnaire, I am reasonably satisfied that the contractor has the competence to carry out or, as the case may be, manage that construction work in compliance with regulation 8(3) of the Construction (Design and Management) Regulations 1994.

Yours faithfully

Planning Supervisor
for and on behalf of

cc. [Contractor]

REQUEST FOR ADDRESSING ADEQUACY OF RESOURCES

[Contractor] Ref:
[Address]

[Date]

Dear Sir

Construction (Design and Management) Regulations 1994: Adequacy of resourcing – regulation 9(3)
Project: . . . .

Further to the appointment of the principal contractor, it is a requirement in compliance with regulation 9(3) that the allocation of adequate resources is established prior to letting the contract.

In this respect I would be pleased if you could contact this office so that arrangements can be made to address the issue.

Yours faithfully

Planning Supervisor
for and on behalf of

cc. [Client]

ALLOCATION OF ADEQUATE RESOURCES: SAMPLE INTERVIEW

MINUTES OF MEETING

Date issued: [date] Minutes prepared by: GJB

Project:	Installation and commissioning of ready-mixed concrete batching plant
Contract No.:	C737
Purpose of meeting:	Adequacy of resourcing – regulation 9(3)
Place of meeting:	[details]
Time and Date:	[date]

Present:	Representing:	Position:
GJB	[details]	Planning Supervisor
AR	[details]	Contract Manager

Date of next meeting: N/A

Distribution:
AR [details]
CH [details]

Notes to recipients:

(1) Departments or individuals are expected to implement any actions as noted.

(2) These Minutes record [details] understanding of the Meeting and intended actions arising therefrom. Your agreement that the Minutes form a true record of the discussions will be assumed unless adverse comments are received in writing within 5 days of receipt of these Minutes.

Contd.

Contd.

MINUTES OF MEETING

(Continuation sheet no. 2)

The following resumé constitutes the minutes of an interview meeting undertaken [date] to establish the adequacy of resourcing in compliance with regulation 9(3) of the Construction (Design and Management) Regulations 1994.

		ACTION
1	**Nature of the project**	
1.1	[Full name of AR] (AR) for and on behalf of [details of construction company] confirmed that adequate time was allowed for tender submission.	
1.2	AR further confirmed that the contract duration period was adequate and that he was fully conversant with the type of work to be undertaken.	AR
2	**Existing environment**	
2.1	AR confirmed full awareness of the working quarry environment and that all staff and operatives would be made aware of the associated health and safety issues through the company's standard induction procedures.	
2.2	AR confirmed that the hazards identified within this section of the pre-tender health and safety plan had all been noted and due account taken.	AR
2.3	The contractor further confirmed that the hazards associated with the vertical rock face, the drainage sump and bituminous coating plant would be subject to risk assessments and method statements which would be included in the construction phase health and safety plan.	AR
3	**Existing drawings**	
3.1	The contractor confirmed that he had received all drawings as listed within the contract documentation.	
4	**The design**	
4.1	The contractor confirmed he fully understood the nature of the design and its implications in terms of stability during the construction sequence.	AR
5	**Construction materials**	
5.1	AR confirmed he had accounted for all items in this section.	
5.2	AR further confirmed that COSHH assessments would form an integral part of the construction phase health and safety plan and would be highlighted during the induction programme.	

Contd.

Contd.

MINUTES OF MEETING

(Continuation sheet no. 3)

	ACTION
6 Site-wide elements	
6.1 AR confirmed that all aspects of this section of the pre-tender health and safety plan had been appraised.	
6.2 The contractor further confirmed that all emergency routes would be kept unobstructed at all times.	AR
6.3 GB informed the contractor that a separate contract would be starting at the same time as [details of contract].	AR
6.4 AR noted the site accommodation limitations but acknowledged that arrangements were adequate.	
7 Overlap with client's undertaking	
7.1 Confirmation was provided by the contractor that he fully understood the implications of this section of the pre-tender health and safety plan.	
7.2 He further confirmed that strict control would be applied for compliance with the quarry manager's rules.	AR
7.3 AR also confirmed that such rules and procedures would receive due emphasis in the induction process.	AR
7.4 The contractor stressed that all of [details of] workforce would be issued with identity badges carrying photographs, which would be worn at all times. A similar arrangement was implemented for sub-contractors.	AR
8 Site rules	
8.1 All site rules were confirmed as standard procedures by the contractor.	AR
9 Continuing liaison	
9.1 AR confirmed that the construction phase health and safety plan would be forwarded 1 week before commencement of construction, and would include specific method statements in relation to:	AR
• Traffic and material management • Site security • Control methods for dust and noise suppression • Liaison with quarry manager • Welfare provision • Emergency procedures.	

Contd.

Contd.

MINUTES OF MEETING

(Continuation sheet no. 4)

		ACTION
9.2	AR confirmed that he understood the resource implication associated with the preparation of the health and safety file. He further confirmed that information for inclusion in this file would be passed over to the planning supervisor as and when it become available.	AR
9.3	The contractor further confirmed that he understood the need for method statements specific to the work being undertaken, and additional method statements would support the development of the construction phase health and safety plan compatible with contract progress.	
9.4	AR confirmed that welfare arrangements would be installed on the first day of construction commencing and would consist of flush toilets, hot and cold water and canteen/mess room facilities.	AR
10	**General statement**	
10.1	AR for and on behalf of [details of construction company] confirmed that they had adequately resourced health and safety management issues as identified within the pre-tender health and safety plan in addition to the obligations which they as a competent contractor would ordinarily resource.	
10.2	AR further confirmed that all the above health and safety resource provisions had been reflected in the tender price submitted.	AR

Signed: . . . .

GB
for and on behalf of
[details]

AR
[details]

Date: . . . .

REQUEST FOR CONTRACTOR'S SIGNATURE TO MINUTES OF ADEQUACY OF RESOURCES MEETING

[Contractor] Ref:
[Address]

[Date]

Dear Sir

Construction (Design and Management) Regulations 1994: Adequacy of resourcing – regulation 9(3)
Project: . . . .

Further to the interview session held yesterday, please find enclosed minutes of the issues addressed corresponding to the establishment of the allocation of adequacy of resources.

I would be pleased if you would read, amend, sign and return as soon as possible so that compliance with regulation 9(3) can be confirmed.

Yours faithfully

Planning Supervisor
for and on behalf of

CONFIRMATION OF ALLOCATION OF ADEQUATE RESOURCES

[Client] Ref:
[Address]

[Date]

Dear Sir

Construction (Design and Management) Regulations 1994: Adequacy of resourcing – regulation 9(3)
Project: . . . .

Further to the specific enquiries made in relation to the above project, I am reasonably satisfied that the contractor has allocated or, as appropriate, will allocate adequate resources to enable the contractor to comply with the requirements and prohibitions imposed on him by or under the relevant statutory provisions, in compliance with regulation 9(3) of the Regulations.

Yours faithfully

Planning Supervisor
for and on behalf of

cc. [Contractor]

REQUEST FOR PRINCIPAL CONTRACTOR TO COMPLETE RELEVANT SECTIONS OF F10 NOTIFICATION [ORIGINAL F10 TO BE AMENDED (PS1)]

[Contractor] Ref:
[Address]

[Date]

Dear Sir

Construction (Design and Management) Regulations 1994: Additional F10 Notification
Project: . . . .

Further to your appointment as Principal Contractor on the above project, I would be pleased if you would complete sections 7(c), 7(d), 9 and 11 of the attached form F10 and return it to the writer for onward transmission to the Health and Safety Executive, as soon as possible.

I also draw your attention to the Principal Contractor's obligations under regulation 16(1)(d) of the Construction (Design and Management) Regulations 1994 to 'ensure that the particulars required to be in any notice given under regulation 7 are displayed in a readable condition in a position where they can be read by any person at work on construction work in connection with the project'.

Yours faithfully

Planning Supervisor
for and on behalf of

Enc. F10

ADDITIONAL F10 NOTIFICATION

Health and Safety Executive Ref:
[Address]

[Date]

Dear Sir

Construction (Design and Management) Regulations 1994: F10 additional notification – regulation 7(4)
Project: . . . .

Please find enclosed form F10 for additional notification of the above project in compliance with regulation 7(4) of the Construction (Design and Management) Regulations 1994.

Yours faithfully

Planning Supervisor
for and on behalf of

Enc. F10 (additional)

cc. [Client]
[Principal Contractor]

F10 DISPLAY

[Contractor] Ref:
[Address]

[Date]

Dear Sir

Construction (Design and Management) Regulations 1994: F10 additional notification
Project: . . . .

Please find enclosed a copy of the additional F10 notification as forwarded to the local Health and Safety Executive office.

I draw your attention to the requirement in compliance with regulation 16(1)(d) of the Construction (Design and Management) Regulations 1994 for you to 'ensure that the particulars required to be in any notice given under regulation 7 are displayed in a readable condition in a position where they can be read by any person at work on construction work in connection with the project'.

Yours faithfully

Planning Supervisor
for and on behalf of

cc. [Client]

REQUEST FOR DRAFT COPY OF CONSTRUCTION PHASE HEALTH AND SAFETY PLAN

[Contractor] Ref:
[Address]

[Date]

Dear Sir

Construction (Design and Management) Regulations 1994: Construction phase Health and Safety plan
Project: . . . .

Further to your appointment as Principal Contractor on the above contract it is a requirement in compliance with regulation 10 of the Construction (Design and Management) Regulations 1994 that the client must ensure that:

'the construction phase of any project does not start unless a health and safety plan complying with regulation 15(4) has been prepared in respect of the project'.

I would therefore be pleased if you could forward the construction phase health and safety plan at your earliest convenience.

Yours faithfully

Planning Supervisor
for and on behalf of

cc. [Client]

SANCTION OF SUITABILITY OF CONSTRUCTION PHASE HEALTH AND SAFETY PLAN

[Client] Ref:
[Address]

[Date]

Dear Sir

Construction (Design and Management) Regulations 1994
Start of construction phase – regulation 10
Project: . . . .

Further to the receipt of the Principal Contractor's construction phase health and safety plan, I am pleased to confirm compliance with regulation 15(4) of the Construction (Design and Management) Regulations 1994 and therefore offer no reason in terms of health and safety management why the construction phase should not commence.

Yours faithfully

Planning Supervisor
for and on behalf of

cc. Principal Contractor

HEALTH AND SAFETY MANAGEMENT AGENDA ITEM: PROGRESS MEETING

Progress meeting: . . . . Project: . . . .
Date: . . . . Venue: . . . .

Health and safety management agenda item:		Yes	No	Regulation
(1)	Is the HSE additional notification (F10) displayed in a prominent position?	☐	☐	16(1)(d)
(2)	Is the construction phase health and safety plan on site and being developed?	☐	☐	15(4)
(3)	Has an Inspector from the Health and Safety Executive visited the site?	☐	☐	
(4)	If yes, please outline any relevant comments:			
(5)	Have any improvement or prohibition notices been issued?	☐	☐	
(6)	If so, please enlarge:			
(7)	Have there been any accidents, injuries and/or dangerous occurrences since the last meeting?	☐	☐	
(8)	If yes, please outline:			
(9)	Have any design-sensitive variation orders been issued in the last month?	☐	☐	
(10)	If so, please identify:			
(11)	Have these been subject to a risk assessment strategy?	☐	☐	13(2)
(12)	Have any traffic management problems been evident on site?	☐	☐	
(13)	Are pedestrian segregation measures effective?	☐	☐	
(14)	Are emergency exit/access routes and assembly points adequately safeguarded?	☐	☐	
(15)	Please outline:			
(16)	Have all employees received adequate induction/ training	☐	☐	16(3), 17(1)

Contd.

Contd.

Health and safety management agenda item:		Yes	No	Regulation
(17)	Have all sub-contractors been assessed for competence?	☐	☐	8(3)
(18)	Are risk assessments/COSHH assessments being undertaken?	☐	☐	
(19)	Is relevant information being collated for the health and safety file?	☐	☐	14(d)
(20)	Do you wish to pass any information to the Planning Supervisor for inclusion in the health and safety file?	☐	☐	16(1)(e)
(21)	Are site security measures adequate to prevent trespass?	☐	☐	16(1)(c)
(22)	Are welfare facilities effective and adequate?	☐	☐	

HEALTH AND SAFETY FILE ACCOMPANYING LETTER AND CONFIRMATION SLIP

[Client] Ref:
[Address]

[Date]

Dear Sir

Construction (Design and Management) Regulations 1994: Health and Safety file – regulation 14(f)
Project: . . . .

Please find enclosed the health and safety file for the above project in compliance with regulation 14(f) of the Construction (Design and Management) Regulations 1994.

This document must be maintained current and available for inspection in compliance with regulation 12(i), and it should be noted that:

'Every client shall take such steps as it is reasonable for a person in his position to take to ensure that the information in any health and safety file which has been delivered to him is kept available for inspection by any person who may need information in the file for the purpose of complying with the requirements and prohibitions imposed on him by or under the relevant statutory provisions.'

Furthermore, any transference of ownership of the property must be accompanied by transference of the corresponding health and safety file in compliance with regulation 12(2):

'It shall be sufficient compliance with paragraph (1) by a client who disposes of his entire interest in the property of the structure if he delivers the health and safety file for the structure to the person who acquires his interest in the property of the structure and ensures such person is aware of the nature and purpose of the health and safety file.'

I would therefore be grateful if you would acknowledge receipt of the health and safety file for the project by signing and returning the attached sheet.

Yours faithfully

Planning Supervisor
for and on behalf of

HEALTH AND SAFETY FILE: CONFIRMATION OF RECEIPT

I acknowledge receipt of the health and safety file for:

[contract]

. . . .

Signed:

Name: . . . . Designation: . . . . Signature: . . . .

for and on behalf of: . . . . Date: . . . .

Please return to: Planning Supervisor
[Address]

SAMPLE CONTENTS OF HEALTH AND SAFETY FILE

The health and safety file should include information about all the following topics, where this may be relevant to the health and safety of any future construction work. The level of detail should be proportionate to the risks likely to be involved in such work.

(1) A brief description of the work carried out.

(2) Residual hazards and how they have been dealt with (for example, surveys or other information concerning asbestos, contaminated land, water bearing strata, buried services).

(3) Key structural principles incorporated in the design of the structure (e.g. bracing, sources of substantial stored energy – including pre or post-tensioned members) and safe working loads for floors and roofs, particularly where these may preclude placing scaffolding or heavy machinery there.

(4) Any hazards associated with the materials used (for example, hazardous substances, lead paint, special coatings which should not be burned off).

(5) Information regarding the removal or dismantling of installed plant and equipment (for example, lifting arrangements).

(6) Health and safety information about equipment provided for cleaning or maintaining the structure.

(7) The nature, location and markings of significant services, including fire fighting services.

(8) Information and 'as built' drawings of the structure, its plant and equipment (e.g. the means of safe access to and from service voids, fire doors and compartmentation).

DESIGNER

EXAMPLE OF RISK ASSESSMENT PROFORMA

PROJECT	DATE
PART OF PROJECT	COMPLETED BY:

Ref	Activity element	Significant potential hazards	Population at risk	Risk classification			Design action to be taken to reduce risk	Design action		Reduced risk factors	Future action
				Likelihood	Severity	Risk rating		By	Date completed		

Risk assessment
Likelihood: remote = 1; possible = 2; probable = 3.
Severity: minor = 1; serious = 2; severe/fatal = 3.
Risk rating: low = 1, 2; medium = 3, 4; high = 6–9.

EXAMPLE OF RISK REGISTER

Ref	Activity element	Significant/ principal potential hazards	Design action to be taken to reduce risk	Current action	Future action
Design team				Contractor responsibility	Planning supervisor and/or facilities management

PRINCIPAL CONTRACTOR

CONSTRUCTION PHASE HEALTH AND SAFETY PLAN: SAMPLE CONTENTS

1. Description of the project	• Project description and programme details • Details of client, designers, planning supervisor and other consultants • Extent and location of existing records and plans
2. Communication and management of the work	• Management structure and responsibilities • Health and safety goals for the project and arrangements for monitoring and review of health and safety performance • Arrangements for: ○ regular liaison between parties on site ○ consultation with the workforce ○ the exchange of design information between the client, designers, planning supervisor and contractors on site • Handling design changes during the project • The selection and control of contractors • The exchange of health and safety information between contractors • Security, site induction and on site training • Welfare facilities and first aid • The reporting and investigation of accidents and incidents including near misses • The production and approval of risk assessments and method statements SITE RULES FIRE AND EMERGENCY PROCEDURES
3. Arrangements for controlling significant site risks	**SAFETY RISKS:** • Services including temporary electrical installations • Preventing falls • Work with or near fragile materials • Control of lifting operations • Dealing with services (water, electricity and gas) • The maintenance of plant and equipment • Poor ground conditions • Traffic routes and segregation of vehicles and pedstrians • Storage of hazardous materials • Dealing with existing unstable structures • Accommodating adjacent land use • Other significant safety risks **HEALTH RISKS:** • Removal of asbestos • Dealing wtih contaminated land • Manual handling • Use of hazardous substances • Reducing noise and vibrations • Other significant health risks
4. The health and safety file	LAYOUT AND FORMAT ARRANGEMENTS FOR THE COLLECTION AND GATHERING OF INFORMATION STORAGE OF INFORMATION

HAZARD IDENTIFICATION CHECKLIST AND METHOD STATEMENT PROPOSAL FORM

HAZARD IDENTIFICATION CHECKLIST (Sheet 1 of 2) **Completed by:** . . . . **Date:**	**Client:** . . . . **Contract:** . . . .

	Yes	No
CONTRACTUAL		
Have you received:		
Relevant aspects of construction phase health and safety plan	☐	☐
Corresponding induction information	☐	☐
EXISTING ENVIRONMENT		
Does work involve:		
Asbestos removal	☐	
Confined spaces	☐	
Contact with:		
Drainage systems	☐	
Sewage systems	☐	
Vermin	☐	
Water supply	☐	
Interruptions to circuits:		
Computer systems	☐	
Fire alarm	☐	
Lightning conductor	☐	
Security	☐	
Smoke detection	☐	
Live circuitry	☐	
Occupied premises	☐	
Ongoing contracts	☐	
Other contract interfaces	☐	
Pedestrian movement	☐	
Removal of pre-1973 luminaires	☐	
Removal of paint systems	☐	
Vehicular management	☐	
Work in pre-1960 buildings	☐	
Is this site sensitive to issues of:		
Dust	☐	
Lighting	☐	
Noise	☐	
PROCEDURES		
Does work involve:		
Access/egress facilities:		
Ladders	☐	
Platforms (hydraulic)	☐	
Restricted Access	☐	
Scaffolding	☐	
Scissor lifts	☐	
Shared Access	☐	
Staging	☐	
Towers	☐	
Confined Spaces	☐	
Live Circuitry	☐	
Occupied Premises	☐	
Work at Depth	☐	

	Yes	No	Trade name
CONSTRUCTION MATERIALS			
Are any of the following to be used?			
Admixtures	☐		
Aluminium sulphate	☐		
Asbestos	☐		
Chlorine	☐		
Cleaning agents	☐		
Damp-proof treatments	☐		
Descaling agents	☐		
Degreasers	☐		
Epoxies	☐		
Fluxes	☐		
Lime	☐		
Man made mineral fibre	☐		
Painting systems	☐		
Solvents	☐		
Sulphuric acid	☐		
Wood preservative treatment	☐		
Other substances (please specify):	☐		
Have hazard data sheets been obtained for all these items ticked above?	☐	☐	
Have associated COSHH assessments been undertaken?	☐	☐	
CONSTRUCTION EQUIPMENT			
Are any of the following to be used?			
Acetylene	☐		
Calor	☐		
Compressors	☐		
Excavators	☐		
Generators	☐		
Hydraulic work	☐		
Lifting	☐		
Liquefied petroleum gas	☐		
Mains	☐		
Oxygen	☐		
Transformers	☐		
Other gas cylinders	☐		
Systems:			
Electrical	☐		
Mechanical	☐		
Pneumatic	☐		
Proprietary	☐		
Support	☐		
Have risk assessments been undertaken?	☐	☐	

Contd.

Contd.

	Yes	No		Yes	No	Trade name
SITE-WIDE ELEMENTS			PROCESS			
Are there difficulties in relation to:			Does work on site involve:			
Access/egress to site	☐		Burning	☐		
Parking limitations	☐		Chasing	☐		
Phasing of the works	☐		Cutting	☐		
Site security	☐		Discing	☐		
Traffic patterns	☐		Generation of fumes	☐		
Trespass/vandalism	☐		Gouging	☐		
			Grinding	☐		
REGISTERS: ARE THE			Manual handling	☐		
FOLLOWING IN PLACE?			Naked flames	☐		
F2508 . . . RIDDOR . . .	☐		Notching	☐		
Reportable Accidents			Sanding	☐		
B1510 . . . . Site Accident	☐		Stripping	☐		
Book			Welding	☐		
Inspection reports Construction			Other processes (please specify):	☐		
(Health, Safety and Welfare)			**Have risk assessments been**	☐	☐	
Regulations 1996			**undertaken for items ticked**			
Excavation	☐		**above?**			
Scaffolding	☐					
Working platforms	☐					

METHOD STATEMENT PROPOSAL FORM

METHOD STATEMENT PROPOSAL FORM	**Client:** . . . . **Contract:** . . . .

Company:
Address:
Tel no.:
Contact name:

WORK ELEMENT
Duration of work: . . . . Starting date: . . . .
Completion date: . . . .

SITE PERSONNEL
Responsible person's name: . . . . Tel No.: . . . .
Operatives' names: 1. . . . . 4. . . . .
2. . . . . 5. . . . .
3. . . . . 6. . . . .
Sub-contractor's name: . . . . Tel No.: . . . .
Address:

	Yes	No
Have all site personnel received appropriate training and education?	☐	☐
Have all sub-contractors been assessed for competence?	☐	☐

PLANT/EQUIPMENT
Please itemise:

MATERIALS
Please itemise:

	Yes	No
Are appropriate hazard sheets attached?	☐	☐

RISK ASSESSMENT

Likelihood:	Remote ☐	Possible ☐	Probable ☐
Severity:	Minor ☐	Serious ☐	Severe/fatal ☐
Risk rating:	Low ☐	Medium ☐	High ☐

METHOD STATEMENT

Note:
Method statement to emphasise the appropriate use of:

Communication	☐	Pedestrian segregation	☐	Site management structure	☐
Emergency exit routes	☐	Permits to work	☐	Supports system	☐
Induction talks	☐	Personal protective equipment	☐	Traffic management	☐
Liaison	☐	Phasing	☐	Welfare arrangements	☐
Material management	☐	Risk assessments	☐		
Means of access/egress	☐	Signage	☐		

CONTRACTOR

HAZARD IDENTIFICATION CHECKLIST AND METHOD STATEMENT PROPOSAL FORM

HAZARD IDENTIFICATION CHECKLIST (Sheet 1 of 2)	**Client:** ....
Completed by: .... **Date:**	**Contract:** ....

	Yes	No		Yes	No	Trade name
CONTRACTUAL			CONSTRUCTION MATERIALS			
Have you received:			Are any of the following to be used?			
Relevant aspects of construction phase health and safety plan	☐	☐	Admixtures	☐		
Corresponding induction information	☐	☐	Aluminium sulphate	☐		
			Asbestos	☐		
EXISTING ENVIRONMENT			Chlorine	☐		
Does work involve:			Cleaning agents	☐		
Asbestos removal	☐		Damp-proof treatments	☐		
Confined spaces	☐		Descaling agents	☐		
Contact with:			Degreasers	☐		
Drainage systems	☐		Epoxies	☐		
Sewage systems	☐		Fluxes	☐		
Vermin	☐		Lime	☐		
Water supply	☐		Man made mineral fibre	☐		
Interruptions to circuits:			Painting systems	☐		
Computer systems	☐		Solvents	☐		
Fire alarm	☐		Sulphuric acid	☐		
Lightning conductor	☐		Wood preservative treatment	☐		
Security	☐		Other substances (please specify):	☐		
Smoke detection	☐					
Live circuitry	☐		Have hazard data sheets been obtained for all these items ticked above?	☐	☐	
Occupied premises	☐					
Ongoing contracts	☐		Have associated COSHH assessments been undertaken?	☐	☐	
Other contract interfaces	☐					
Pedestrian movement	☐		CONSTRUCTION EQUIPMENT			
Removal of pre-1973 luminaires	☐		Are any of the following to be used?			
Removal of paint systems	☐		Acetylene	☐		
Vehicular management	☐		Calor	☐		
Work in pre-1960 buildings	☐		Compressors	☐		
			Excavators	☐		
Is this site sensitive to issues of:			Generators	☐		
Dust	☐		Hydraulic work	☐		
Lighting	☐		Lifting	☐		
Noise	☐		Liquefied petroleum gas	☐		
			Mains	☐		
PROCEDURES			Oxygen	☐		
Does work involve:			Transformers	☐		
Access/egress facilities:			Other gas cylinders	☐		
Ladders	☐					
Platforms (hydraulic)	☐		Systems:			
Restricted Access	☐		Electrical	☐		
Scaffolding	☐		Mechanical	☐		
Scissor lifts	☐		Pneumatic	☐		
Shared Access	☐		Proprietary	☐		
Staging	☐		Support	☐		
Towers	☐					
Confined Spaces	☐		**Have risk assessments been undertaken?**	☐	☐	
Live Circuitry	☐					
Occupied Premises	☐					
Work at Depth	☐					

Contd.

Contd.

	Yes	No		Yes	No	Trade name
SITE-WIDE ELEMENTS			PROCESS			
Are there difficulties in relation to:			Does work on site involve:			
Access/egress to site	☐		Burning	☐		
Parking limitations	☐		Chasing	☐		
Phasing of the works	☐		Cutting	☐		
Site security	☐		Discing	☐		
Traffic patterns	☐		Generation of fumes	☐		
Trespass/vandalism	☐		Gouging	☐		
			Grinding	☐		
REGISTERS: ARE THE FOLLOWING IN PLACE?			Manual handling	☐		
			Naked flames	☐		
F2508 . . . RIDDOR . . . Reportable Accidents	☐		Notching	☐		
			Sanding	☐		
B1510 . . . . Site Accident Book	☐		Stripping	☐		
			Welding	☐		
Inspection reports Construction (Health, Safety and Welfare) Regulations 1996			Other processes (please specify):	☐		
			Have risk assessments been undertaken for items ticked above?	☐	☐	
Excavation	☐					
Scaffolding	☐					
Working platforms	☐					

METHOD STATEMENT PROPOSAL FORM

METHOD STATEMENT PROPOSAL FORM	**Client:** . . . . **Contract:** . . . .

Company:
Address:
Tel no.:
Contact name:

WORK ELEMENT
Duration of work: . . . . Starting date: . . . .
Completion date: . . . .

SITE PERSONNEL
Responsible person's name: . . . . Tel No.: . . . .
Operatives' names: 1. . . . . 4. . . . .
2. . . . . 5. . . . .
3. . . . . 6. . . . .
Sub-contractor's name: . . . . Tel No.: . . . .
Address:

	Yes	No
Have all site personnel received appropriate training and education?	☐	☐
Have all sub-contractors been assessed for competence?	☐	☐

PLANT/EQUIPMENT
Please itemise:

MATERIALS
Please itemise:

	Yes	No
Are appropriate hazard sheets attached?	☐	☐

RISK ASSESSMENT

Likelihood:	Remote ☐	Possible ☐	Probable ☐
Severity:	Minor ☐	Serious ☐	Severe/fatal ☐
Risk rating:	Low ☐	Medium ☐	High ☐

METHOD STATEMENT

Note:
Method statement to emphasise the appropriate use of:

Communication	☐	Pedestrian segregation	☐	Site management structure	☐
Emergency exit routes	☐	Permits to work	☐	Supports system	☐
Induction talks	☐	Personal protective equipment	☐	Traffic management	☐
Liaison	☐	Phasing	☐	Welfare arrangements	☐
Material management	☐	Risk assessments	☐		
Means of access/egress	☐	Signage	☐		

APPENDICES

A COMPLETED DESIGNER(S) RISK ASSESSMENT PROFORMA

The designer's risk assessment proforma represents a key document which, in conjunction with other evidence related to the design process, such as minutes of briefing meetings and design workshops, enables a documentation trail to be established compatible with an appropriate design risk strategy. This is essential in the discharge of designer's duties and helps to demonstrate compliance with regulation 13 of the CDM Regulations.

The proforma is an enabling document which not only provides evidence of the design contribution to health and safety management but also communicates to others the residual health and safety related risks pertinent to the project.

The health and safety aspects of design strategy focus on significant and principal hazards specific to the project with no requirement for the consideration of all foreseeable risks. The design response is based on the competent contractor and it is not for the designer to function as the contractor's alter ego in this respect.

Compliance with regulation 13 is initiated by the identification of health and safety hazards, and a well-presented risk assessment proforma is of little value at best and dangerous and misleading at worst if the designer does not possess the competence to identify the underlying hazard initially. Good design practice acknowledges that hazard identification and risk appraisal are only two parts of the risk assessment strategy. The real challenge to design lies in the response to that appraisal through a contribution to health and safety management based on a hierarchy of:

- Elimination
- Reduction/minimisation
- Transfer

The latter option is only exercisable after considerations of the other two have been sensibly exhausted.

The designer's response is governed by 'as far as is reasonably practicable' and thus the contribution must not be disproportionate. However, health and safety, in conjunction with design aspects of form, function, fitness of purpose, aesthetics, environmental factors and cost, must all receive due consideration.

There are no unique proforma models being used unilaterally by the design fraternity and, whilst compliance with regulation 13 represents an absolute duty, the manner in which this is achieved remains judgmental. The response and documentation need to be appropriate and for smaller projects a simple statement on a drawing may suffice. However, the risk assessment proforma can itself provide evidence of a systematic approach and the more effective forms are able to demonstrate a contribution to health and safety management as well as fulfilling the role of a communication device for the transference of residual risks to others. Such forms are therefore vital to the function of the planning supervisor in the drafting of both the pre-tender health and safety plan and the health and safety file, as well as being instructive to other designers in a multi-disciplinary design team.

Whilst many risk assessment techniques are available, the simplest is often the most effective and the risk classification based on the product of likelihood and severity on a scale of 1 to 3 provides an ample demonstration of a systematic approach. A wider classification can simply introduce an unwarranted refinement.

In innovative design a series of risk assessment proformas will record the strategy compatible with the numerous stages in the process, whilst for conventional design projects one set of risk assessment proformas will provide a satisfactory record. The strategy must not dominate the design process since its integration is meant to provide a bench mark of good design practice serving the needs of the designer in making a proactive contribution to health and safety management.

The examples overleaf merely demonstrate typical areas where designers can contribute and acknowledge that the completed risk assessment proforma is a summary of the design process to date in compliance with regulation 13 of the CDM Regulations.

Risk classification

Likelihood: Remote = 1 Possible = 2 Probable = 3
Severity: Minor = 1 Serious = 2 Severe/Fatal = 3

Risk rating: 1, 2 = Low; 3, 4 = Medium; 6–9 = High

Project Example			**Address** Insight House, London				**File** 123.	**Date** Sept. 98			
Project element Various			**Designer** A.B. Consultants								
Ref	**Activity element**	**Significant potential hazards**	**Population at risk**	**Risk classification**			**Design action to be taken to reduce risk**	**Design action**		**Reduced risk factor**	**Future action**
				Likelihood	**Severity**	**Rating**		**By**	**Date**		
	Sub-structure:	Deep excavation	Operatives	3	2	6	Raise pile cap level	L.D.	1.9.98	2 × 2 = 4	P.S.
	Pile cap	adjacent to river									H&S plan
	Utilities:	Ground contamination:	Operatives	3	3	9	• Reduce footprint:	E.K.	2.9.98	1 × 3 = 3	P.S.
	Service trench	arsenic	Public				• Re-alignment				H&S file
							• Minimise excavation				
	Maintenance:										
	Pipebridge	Work at height	Maintenance	2	3	6	Use enhanced	L.C.	2.9.98	1 × 3 = 3	PS
							protective systems				H & S file specification
	Initial risk classification must be undertaken before any mitigation has been introduced										
	Design action invites designer to contribute proactively to health & safety management										
	Accountability through dated signature										
	Design contribution could eliminate or reduce risk either by a reduction in likelihood or severity factors, or both										
	All residual risks need to be communicated to planning supervisor and/or designer and/or contractor										

B SYSTEMS APPROACH TO THE HEALTH AND SAFETY PLAN

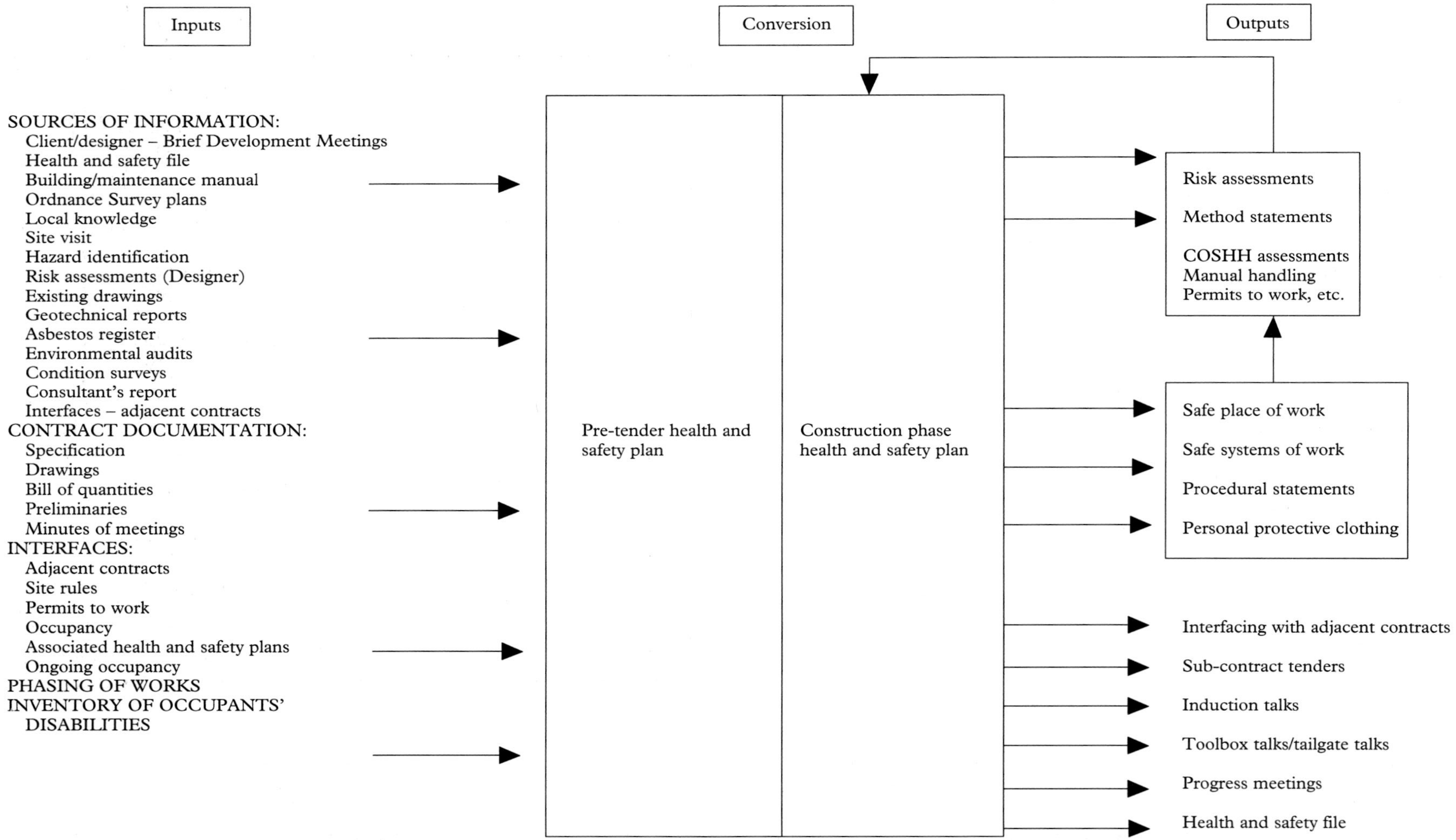
Inputs
Conversion
Outputs
SOURCES OF INFORMATION:
Client/designer – Brief Development Meetings
Health and safety file
Building/maintenance manual
Ordnance Survey plans
Local knowledge
Site visit
Hazard identification
Risk assessments (Designer)
Existing drawings
Geotechnical reports
Asbestos register
Environmental audits
Condition surveys
Consultant's report
Interfaces – adjacent contracts
CONTRACT DOCUMENTATION:
Specification
Drawings
Bill of quantities
Preliminaries
Minutes of meetings
INTERFACES:
Adjacent contracts
Site rules
Permits to work
Occupancy
Associated health and safety plans
Ongoing occupancy
PHASING OF WORKS
INVENTORY OF OCCUPANTS' DISABILITIES
Pre-tender health and safety plan
Construction phase health and safety plan
Risk assessments
Method statements
COSHH assessments
Manual handling
Permits to work, etc.
Safe place of work
Safe systems of work
Procedural statements
Personal protective clothing
Interfacing with adjacent contracts
Sub-contract tenders
Induction talks
Toolbox talks/tailgate talks
Progress meetings
Health and safety file

C SYSTEMS APPROACH TO THE HEALTH AND SAFETY FILE

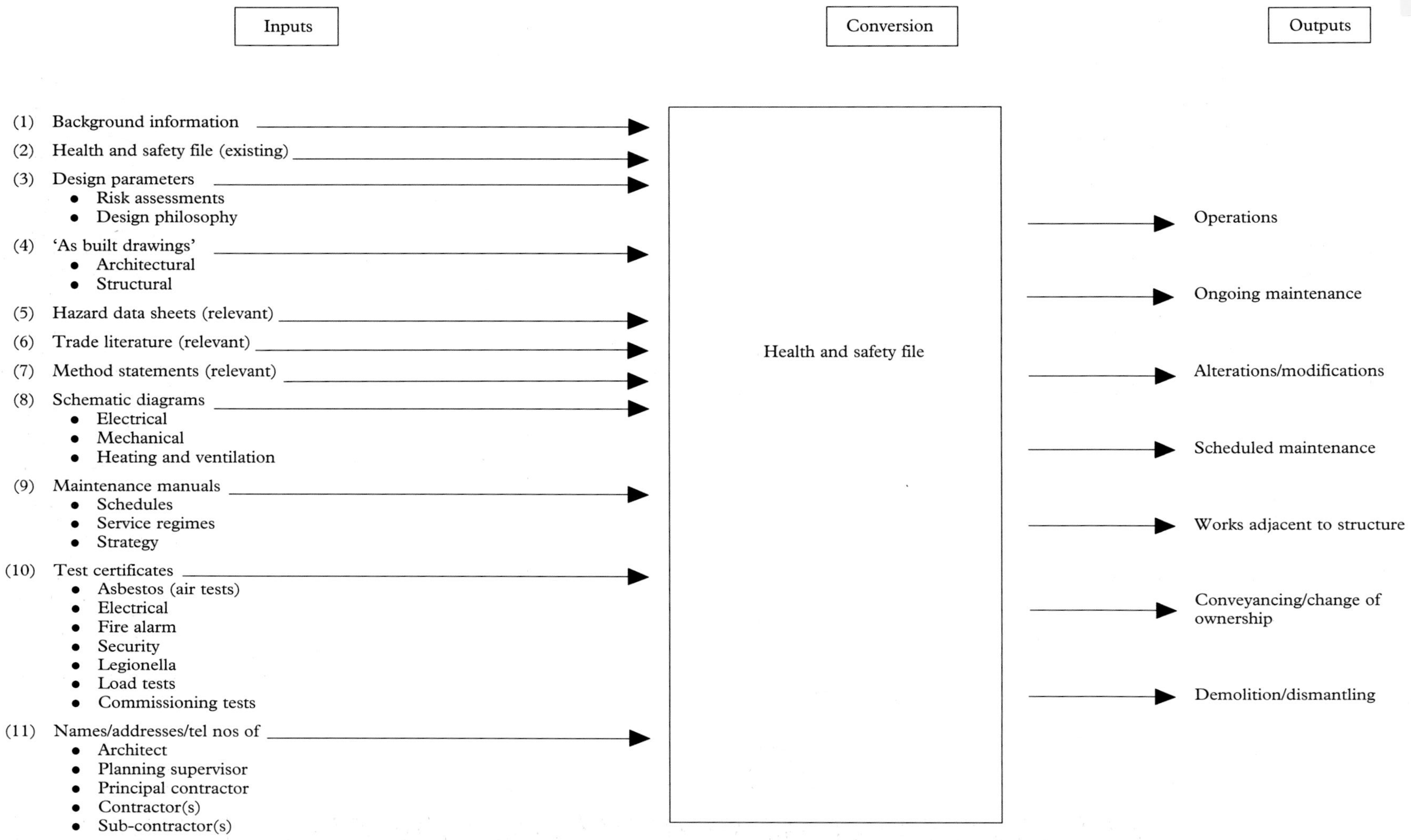
Inputs
Conversion
Outputs
(1) Background information
(2) Health and safety file (existing)
(3) Design parameters
• Risk assessments
• Design philosophy
(4) 'As built drawings'
• Architectural
• Structural
(5) Hazard data sheets (relevant)
(6) Trade literature (relevant)
(7) Method statements (relevant)
(8) Schematic diagrams
• Electrical
• Mechanical
• Heating and ventilation
(9) Maintenance manuals
• Schedules
• Service regimes
• Strategy
(10) Test certificates
• Asbestos (air tests)
• Electrical
• Fire alarm
• Security
• Legionella
• Load tests
• Commissioning tests
(11) Names/addresses/tel nos of
• Architect
• Planning supervisor
• Principal contractor
• Contractor(s)
• Sub-contractor(s)
Health and safety file
Operations
Ongoing maintenance
Alterations/modifications
Scheduled maintenance
Works adjacent to structure
Conveyancing/change of ownership
Demolition/dismantling

D COMPLETED CONTRACTOR'S HAZARD IDENTIFICATION CHECKLIST, RISK ASSESSMENT AND METHOD STATEMENT

The hazard identification checklist, risk assessment and method statement are sequences to be undertaken by the contractor in complying with statutory obligations set out in regulation 3 of the Management of Health and Safety at Work Regulations 1992, namely:

> 'Every employer shall make a suitable and sufficient assessment of:
>
> (a) The risks to the health and safety of his employees to which they are exposed whilst they are at work; and
>
> (b) The risks to the health and safety of persons not in his employment arising out of or in connection with the conduct by him of his undertaking.'

The method statement itself is a detailed procedural statement of intent in response to the risk assessment arising out of the hazard identification exercise. Its purpose is to outline explicitly the methodology to be employed, measures to be taken, plant to be used in providing SAFE SYSTEMS OF WORK and A SAFE PLACE OF WORK in carrying out the intended operation in a suitable and sufficient manner.

Such method statements facilitate the management of risk and for CDM-related projects will contribute to the development of the construction phase health and safety plan. Each method statement must be operation specific and, whilst checklists can help as an aide memoire, the written statement must account for all salient factors applicable to that operation. A generic approach does not deal with specifics and can fail to identify the one factor making this operation more hazardous.

The two forms overleaf develop the process based on accountability and enable a unique method statement to be drafted prior to the commencement of the operation, through a checklist approach. This appendix is presented as an example for use on standard construction operations. Projects with particularly hazardous operations would require a more involved approach with an input from a number of competent personnel, and the approach outlined would need to be expanded further.

HAZARD IDENTIFICATION CHECKLIST (Sheet 1 of 2) **Completed by:** SDS **Date:** 6.11.98	**Client:** AB LTD. Sheet 1 of 2 **Contract:** Factory extension

	Yes	No
CONTRACTUAL		
Have you received:		
Relevant aspects of construction phase health and safety plan	☑	☐
Corresponding induction information	☑	☐
EXISTING ENVIRONMENT		
Does work involve:		
Asbestos removal	☐	
Confined spaces	☑	
Contact with:		
Drainage systems	☐	
Sewage systems	☑	
Vermin	☐	
Water supply	☐	
Interruptions to circuits:		
Computer systems	☐	
Fire alarm	☐	
Lightning conductor	☐	
Security	☐	
Smoke detection	☐	
Live circuitry	☐	
Occupied premises	☐	
Ongoing contracts	☐	
Other contract interfaces	☐	
Pedestrian movement	☐	
Removal of pre-1973 luminaires	☐	
Removal of paint systems	☐	
Vehicular management	☐	
Work in pre-1960 buildings	☐	
Is this site sensitive to issues of:		
Dust	☐	
Lighting	☐	
Noise	☐	
PROCEDURES		
Does work involve:		
Access/egress facilities:		
Ladders	☑	
Platforms (hydraulic)	☐	
Restricted Access	☐	
Scaffolding	☐	
Scissor lifts	☐	
Shared Access	☐	
Staging	☐	
Towers	☐	
Confined Spaces	☑	
Live Circuitry	☐	
Occupied Premises	☑	
Work at Depth	☐	

	Yes	No	Trade name
CONSTRUCTION MATERIALS			
Are any of the following to be used?			
Admixtures	☐		
Aluminium sulphate	☐		
Asbestos	☐		
Chlorine	☐		
Cleaning agents	☐		
Damp-proof treatments	☐		
Descaling agents	☐		
Degreasers	☐		
Epoxies	☐		
Fluxes	☐		
Lime	☐		
Man made mineral fibre	☐		
Painting systems	☐		
Solvents	☐		
Sulphuric acid	☐		
Wood preservative treatment	☐		
Other substances (please specify):	☐		
Have hazard data sheets been obtained for all these items ticked above?	☐	☐	
Have associated COSHH assessments been undertaken?	☐	☐	
CONSTRUCTION EQUIPMENT			
Are any of the following to be used?			
Acetylene	☐		
Calor	☐		
Compressors	☑		
Excavators	☑		
Generators	☑		
Hydraulic work	☐		
Lifting	☐		
Liquefied petroleum gas	☐		
Mains	☐		
Oxygen	☐		
Transformers	☐		
Other gas cylinders	☐		
Systems:			
Electrical	☐		
Mechanical	☐		
Pneumatic	☐		
Proprietary	☐		
Support	☐		
Have risk assessments been undertaken?	☐	☐	

Contd.

Contd.

	Yes	No		Yes	No	Trade name
SITE-WIDE ELEMENTS			PROCESS			
Are there difficulties in relation to:			Does work on site involve:			
Access/egress to site	☑		Burning	☐		
Parking limitations	☐		Chasing	☑		
Phasing of the works	☐		Cutting	☐		
Site security	☐		Discing	☐		
Traffic patterns	☐		Generation of fumes	☐		
Trespass/vandalism	☐		Gouging	☐		
			Grinding	☐		
REGISTERS: ARE THE			Manual handling	☑		
FOLLOWING IN PLACE?			Naked flames	☐		
F2508 . . . RIDDOR . . .	☑		Notching	☐		
Reportable Accidents			Sanding	☐		
B1510 . . . . Site Accident	☑		Stripping	☐		
Book			Welding	☐		
Inspection reports Construction			Other processes (please specify):	☐		
(Health, Safety and Welfare)			**Have risk assessments been**	☐	☐	
Regulations 1996			**undertaken for items ticked**			
Excavation	☐		**above?**			
Scaffolding	☐					
Working platforms	☐					

METHOD STATEMENT PROPOSAL FORM	**Client:** AB Ltd **Contract:** Factory extension

Company: BC Subcontractors Ltd
Address: Main Road, Newtown
Tel no.:
Contact name: LE Davies

WORK ELEMENT Drainage sump
Duration of work: 1 week
Starting date: 6.11.98
Completion date: 12.11.98

SITE PERSONNEL
Responsible person's name: J. Perfect — Tel No.: ....
Operatives' names:
1. A. Wright
2. B. Jones
3. C. Smith
4. ....
5. ....
6. ....

Sub-contractor's name: As above — Tel No.: ....
Address:

	Yes	No
Have all site personnel received appropriate training and education?	☑	☐
Have all sub-contractors been assessed for competence?	☑	☐

PLANT/EQUIPMENT
Please itemise: Hydraulic excavator
2 No 100mm ∅ Pumps
Ladder
Air testing equipment

MATERIALS
Please itemise: Poling boards (32mm thick)
Hydraulic struts
Timber wedges

	Not applicable	
	Yes	No
Are appropriate hazard sheets attached?	☑	☐

RISK ASSESSMENT

Likelihood:	Remote ☐	Possible ☐	Probable ☑
Severity:	Minor ☐	Serious ☐	Severe/fatal ☑
Risk rating:	Low ☐	Medium ☐	High ☑

METHOD STATEMENT
The excavation shall be subject to a 'Permit to Work' control system with air quality tests prior to entry. Ladder to be provided and excavation to be supported by poling boards and struts compatible with advancing face of excavation. Drainage pump to be provided with additional standby pump. Top of excavation to be physically barriered and excavation to be checked prior to entry and at the start of the working day. Personal standards of hygiene to be emphasised. Material surcharge around excavation to be avoided.

Note:
Method statement to emphasise the appropriate use of:

Communication	☐	Pedestrian segregation	☐	Site management structure	☐
Emergency exit routes	☐	Permits to work	☑	Supports system	☑
Induction talks	☑	Personal protective equipment	☑	Traffic management	☐
Liaison	☑	Phasing	☐	Welfare arrangements	☑
Material management	☑	Risk assessments	☐		
Means of access/egress	☑	Signage	☑		

BIBLIOGRAPHY

Anon. (1995) *Management of Construction Safety*. Croner Publications, Kingston-upon-Thames.

Chartered Institue of Building (1992) *Code of Practice for Project Management and Construction and Development*. Chartered Institute of Building, Ascot.

Centre of Construction Law and Management/CIRIA (1994) *Risk management and procurement in construction*. Seventh Annual Conference, London, Sept 1994. CIRIA, London.

CIRIA (1993) *A guide to the control of substances hazardous to health in design and construction*. CIRIA Report 125. CIRIA, London.

CIRIA (1995) *CDM Regulations – case study guidance for designers: an interim report*. CIRIA Report 145. CIRIA, London.

CIRIA (1997) *Experiences of CDM*. CIRIA Report 171. CIRIA, London.

CIRIA (1998) *CDM Regulations – Work sector guidance for designers*. CIRIA Report 166. CIRIA, London.

CIRIA (1998) *CDM Regulations: practical guidance for clients and clients' agents*. CIRIA Report 172. CIRIA, London.

CIRIA (1998) *CDM Regulations: practical guidance for planning supervisors*. CIRIA Report 173. CIRIA, London.

Hambly, E.C. and Hambly, E.A. (1994) *Risk evaluation and realism*. Proceedings of a conference of the Institution of Civil Engineers, London, May 1994, paper 102, pp. 64–71.

Health and Safety Commission (1992) *Management of Health and Safety at Work Regulations 1992*. Approved Code of Practice. Health and Safety Commision, London.

Health and Safety Executive (1995) *Construction (Design and Management) Regulations 1994 – brief for a designer's handbook*. Health and Safety Commission Contract Research Report 71. HSE, London.

Health and Safety Executive (1995) *Information on Site Safety for Designers of Small Building Projects*. Health and Safety Executive Contract Research Report 72. HSE, London.

Health and Safety Executive (1996) *Construction (Design and Management) Regulation 1994, Approved Code of Practice*. HSE, London.

Health and Safety Commission (2001) *Approved Code of Practice, Managing Health and Safety in Construction* (HSG 224). HSE, London.

HMSO (1995) *Construction (Design and Management) Regulations 1994*. HMSO Statutory Instruments 3140. HMSO, London.

INDEX

Please also see the Contents for a comprehensive list of subjects covered in this book